Progress in Mathematics
Volume 80

Topological Methods in Algebraic Transformation Groups

Proceedings of a Conference at Rutgers University

Edited by
Hanspeter Kraft
Ted Petrie
Gerald W. Schwarz

1989
Birkhäuser
Boston · Basel · Berlin

Hanspeter Kraft
Mathematisches Institut
Universität Basel
Rheinsprung 21
CH-4051 Basel
Switzerland

Ted Petrie
Department of Mathematics
Rutgers University
New Brunswick, New Jersey 07102
U.S.A

Gerald W. Schwarz
Department of Mathematics
Brandeis University
Waltham, Massachusetts 02254-9110
U.S.A.

ISBN-13: 978-1-4612-8219-8 e-ISBN-13: 978-1-4612-3702-0

DOI: 10.1007/978-1-4612-3702-0

Library of Congress Cataloging-in-Publication Data
Topological methods in algebraic transformation group / [edited by]
Hanspeter Kraft, Ted Petrie, Gerald W. Schwarz.
p. cm. -- (Progress in mathematics ; v. 80)
Papers from the conference, "Topological Methods in Algebraic Transformation Groups" held at Rutgers University, 4-8 April, 1988.
1. Transformation groups--Congresses. 2. Algebraic topology--Congresses. 3. Geometry, Algebraic--Congresses. I. Kraft, Hanspeter, 1944- . II. Petrie, Ted, 1939- . III. Schwarz, Gerald W., 1946- . IV. Series: Progress in mathematics (Boston, Mass.) ; vol. 80
QA385.T67 1989
514'.2--dc20

Printed on acid-free paper.

Softcover reprint of the hardcover 1st edition 1989

Text prepared by the editors in camera-ready form.

9 8 7 6 5 4 3 2 1

PREFACE

In recent years, there has been increasing interest and activity in the area of group actions on affine and projective algebraic varieties. Techniques from various branches of mathematics have been important for this study, especially those coming from the well-developed theory of smooth compact transformation groups. It was timely to have an interdisciplinary meeting on these topics.

We organized the conference "Topological Methods in Algebraic Transformation Groups," which was held at Rutgers University, 4-8 April, 1988. Our aim was to facilitate an exchange of ideas and techniques among mathematicians studying compact smooth transformation groups, algebraic transformation groups and related issues in algebraic and analytic geometry. The meeting was well attended, and these Proceedings offer a larger audience the opportunity to benefit from the excellent survey and specialized talks presented. The main topics concerned various aspects of group actions, algebraic quotients, homogeneous spaces and their compactifications.

The meeting was made possible by support from Rutgers University and the National Science Foundation. We express our deep appreciation for this support. We also thank Annette Neuen for her assistance with the technical preparation of these Proceedings.

The Editors.

TABLE OF CONTENTS

INTRODUCTION

The subject of smooth transformation groups has been strongly influenced by the following two central problems:

- **Smooth Linearization Problem:** *Is every smooth action of a compact Lie group on Euclidean space conjugate to a linear action?*
- **Smooth Fixed Point Problem:** *Does every smooth action of a compact Lie group on Euclidean space have a fixed point?*

The tools used to settle these problems—and the further problems they generated—have been important themes in the field. During the last few years researchers from algebraic transformation groups, as well as those from smooth transformation groups, have recognized the interest and importance of the analogues of these questions in the algebraic category:

- **Algebraic Linearization Problem:** *Is every algebraic action of a reductive group on affine space conjugate to a linear action?*
- **Algebraic Fixed Point Problem:** *Does every algebraic action of a reductive group on affine space have a fixed point?*

The intention of the Rutgers Conference on "Topological Methods in Algebraic Transformation Groups" was to facilitate an exchange of ideas among mathematicians whose expertise includes smooth, analytic or algebraic transformation groups. The focus was on the two problems above. Experience from smooth transformation groups shows that their study requires a wide range of techniques and topics:

- Group actions on affine space.
- Characterization and structure of affine space.
- Prehomogeneous varieties.
- Orbit spaces and quotients.
- G-vector bundles and K-theory.

Below we discuss the connection of the topics to the conference theme and to the papers in this volume.

Since the time of the conference SCHWARZ has constructed the first non-linearizable actions of reductive groups on affine spaces, giving a negative answer to the Algebraic Linearization Problem. The Algebraic Fixed Point Problem is still open.

Group actions on affine space. The survey of KRAFT provides a broad introduction to the current problems in the theory of algebraic transformation groups. Emphasis is on the case of reductive group actions on afine space. It is a good place to get an overview of the subject.

SNOW's survey is about actions of unipotent groups. These groups are, in some sense, those furthest removed from being reductive. An important question concerns the structure of free **C**-actions on $\mathbf{C}^n$. Is every such action isomorphic to translation by a fixed vector? SNOW reviews both positive results and recent counterexamples.

The paper of DOVERMANN-MASUDA-PETRIE shows how to construct fixed point free real algebraic actions of the icosahedral group on real varieties diffeomorphic to $\mathbf{R}^n$. These are algebraic versions of previous (purely C^∞) examples.

Characterization and structure of affine space. We begin by mentioning the following simple characterization of Euclidean space of dimension at least 5 in the smooth category. It is essential in treating the Smooth Linearization Problem.

Theorem. *For n at least 5, real Euclidean n-space is the only smooth contractible manifold which is simply connected at infinity.*

This theorem can be used to construct a fixed point free smooth action of the icosahedral group on Euclidean space (see BREDON, "Introduction to Compact Transformation Groups").

In the algebraic setting, the analogous problem is to characterize $\mathbf{C}^n$ among algebraic varieties. SUGIE's paper is concerned with the cases $n = 2$ and $n = 3$. He reviews topological and algebraic results and techniques. He discusses applications and related problems (e.g., Zariski's Cancellation Problem).

The paper of TOMDIECK-PETRIE deals with homology planes, i.e., non-singular, acyclic, affine surfaces over **C**. They show how to construct all such surfaces starting from certain configurations of curves in the projective plane. They make progress towards proving the following conjecture of PETRIE:

Conjecture. *The only homology plane with a non-trivial algebraic automorphism group is $\mathbf{C}^2$.*

A positive answer to this conjecture would imply that every $\mathbf{C}^*$-action on $\mathbf{C}^3$ is linearizable (see the survey of KRAFT).

NEUMANN studies the topological classification of curves which are embedded in complex surfaces, and, in particular, curves embedded in $\mathbf{C}^2$. He determines the list of isomorphism types for embedded curves of low genus.

Prehomogeneous varieties. The articles of WINKELMANN, BRION and LUNA-MOSER-VUST concern the structure of homogeneous and prehomogeneous varieties: A G-variety is called prehomogeneous if it contains a dense orbit. Prehomogeneous varieties are those whose structure one should try to understand first. BRION gives a survey and introduction to spherical varieties: A normal G-variety X, where G is a reductive algebraic group, is said to be spherical if a Borel subgroup of G has a dense orbit. Spherical varieties include compactifications of symmetric spaces. BRION points out many of the beautiful properties and uses of these varieties.

The paper of LUNA-MOSER-VUST concerns the structure of spaces which are prehomogeneous with respect to an action of $\mathrm{PSL}_2(\mathbf{C})$. It shows that—somewhat surprisingly—there are examples which are Artin-Moišezon but not algebraic. Artin-Moišezon spaces already arise as quotients of non-affine varieties by finite group actions.

WINKELMANN reviews his classification of all homogeneous complex manifolds of dimension 3 or less. There are examples which are homogeneous with respect to a real Lie group, but not with respect to any complex Lie group.

Orbit spaces and quotients. In order to study the action of an algebraic group on a variety, it is important to be able to form a quotient and study its properties. The survey article of SCHWARZ reviews many of the important topics in the theory of algebraic quotients of affine G-varieties. He reproves the basic results of LUNA, NEEMAN and KRAFT-PETRIE-RANDALL on the algebraic, analytic and topological structure of quotients.

SHEPHERD-BARRON's survey treats the following question: When is the quotient variety rational? That is, when is its function field a purely transcendental extension of $\mathbf{C}$? SHEPHERD-BARRON gives examples of various techniques and their application to classical moduli problems.

G-vector bundles and K-theory. There are several important results known about the theory of algebraic G-vector bundles over G-varieties. We concentrate on the case that the base is a representation of G. Results of BASS-HABOUSH show that all such G-vector bundles are stably trivial, and results of KRAFT show that all such bundles are equivariantly locally trivial (in the Zariski topology). Surprisingly, the recent counterexamples of SCHWARZ to the Algebraic Linearization Problem arise from non-trivial G-vector bundles over representations (see KRAFT's survey).

BASS's paper concerns the problem of characterizing G-vector bundles among G-varieties X over a given variety S. If X is flat over S and every fiber is isomorphic to a representation of G, then one would like to conclude that X is isomorphic to a G-vector bundle over S. BASS shows that this is "stably" true, i.e., $X \times V$ is a G-vector bundle over S for some G-representation V.

LINEARIZING FLAT FAMILIES OF REDUCTIVE GROUP REPRESENTATIONS

HYMAN BASS

Introduction

Let G be a reductive group over $\mathbf{C}$. Let $X \to S$ be a G-morphism of G-varieties where G acts trivially on S. Then for $p \in S$ the fiber $X(p)$ over p is a G-variety, so we can view X as an S-parametrized family $(X(p))_{p\in S}$ of G-varieties. We ask here, under various linearity assumptions on the individual $X(p)$'s, what kind of global linearity properties X must have over S. For example, assuming that X is flat over S and that each $X(p)$ is isomorphic to a G-representation space, one would like to conclude that X is isomorphic to a G-vector bundle over S. The results announced here imply that this is "stably" true, i.e. $X \times V$ is a G-vector bundle over S for some G-representation space V.

We give basic definitions in section 1, and statements of some of our main results in section 2. Details will appear in [B].

1 Asanuma schemes and their properties

(1.1) Terminology. Let G be a reductive algebraic group over $\mathbf{C}$. By a *G-module* or a *G-representation space* we mean a finite dimensional complex vector space V together with a rational representation of G on V. By *G-scheme* we mean a noetherian affine scheme $Y = \operatorname{Spec}(A)$ over $\mathbf{C}$ on which G acts algebraically. Then by a *G-Y-scheme* we mean a G-scheme X equipped with a G-morphism $X \to Y$ of finite type.

Definition. We then call X

- *linear* if, for some G-module V, $X \cong Y \times V$ (isomorphism of G-Y-schemes);

- *stably linear* if, for some G-module W, $X \times W$ is linear; a
- *vector bundle* if $X \to Y$ is isomorphic to a G- vector bundle over Y; and
- *Asanuma* if, for some G-vector bundle $E \to X$, E is a linear G-Y-scheme.

Clearly we have the following implications:

$$\begin{array}{ccc} \text{linear} & \Longrightarrow & \text{stably linear} \\ \Downarrow & & \Downarrow \\ \text{vector bundle} & \Longrightarrow & \text{Asanuma} \end{array}$$

A G-vector bundle $E \to X$ is called *trivial* if, for some G-module V, $E \cong X \times V$ (isomorphism of G-vector bundles over X). It is called *stably trivial* if $E \oplus T$ is trivial for some trivial G-vector bundle T over X.

We now record some basic facts from [**B**], section 2.

(1.2) *Let X be an Asanuma G-Y-scheme. Then its relative tangent bundle $T_{X/Y}$ is a G-vector bundle over X. Moreover,*

$$X \text{ is stably linear} \iff T_{X/Y} \text{ is stably trivial}\ .$$

(1.3) *Let X be a G-Y-scheme and $E \to X$ a G-vector bundle over X. Then, as G-Y-schemes,*

$$X \text{ is Asanuma} \iff E \text{ is Asanuma.}$$

It follows that Asanuma G-Y-schemes form the smallest class of G-Y-schemes containing Y and satisfying (1.3).

(1.4) *Assume that Y is non-singular. If X is an Asanuma G-Y-scheme, then X is non-singular, and each G-vector bundle over X is stably isomorphic to the pullback of a G-vector bundle over Y.*
This is an easy consequence of the results in [**BH**].

From (1.4) and (1.2) we conclude:

(1.5) *Assume that Y is non-singular and that all G-vector bundles over Y are stably trivial (e.g. that Y is a point, or, more generally, a G-module). Then Asanuma G-Y-schemes have the same properties, and they are stably linear.*

2 Flat families

(2.1) Localization and fibers. Let

$$Y = \mathrm{Spec}(A) \longrightarrow S = \mathrm{Spec}(R)$$

be a morphism of affine schemes, corresponding to a ring homomorphism $R \to A$. A point $p \in S$ is a prime ideal of R, to which we associate the local ring R_p and residue class field $R(p) = R_p/pR_p$. Putting $A_p = R_p \otimes_R A$ and $A(p) = R(p) \otimes_R A$, we call $Y_p = \mathrm{Spec}(A_p)$ the *localization* of Y at p, and $Y(p) = \mathrm{Spec}(A(p))$ the *fiber* of Y over p. We say that Y is *flat* over S if A is a flat R-module.

(2.2) We say that $Y \to S$ satisfies "*going down*" if, for every irreducible closed $Z \subset S$, Z is dominated by every irreducible component of its inverse image in Y. This is automatic if Y is faithfully flat over S (cf. [M]).

(2.3) The setting. We fix a commutative diagram

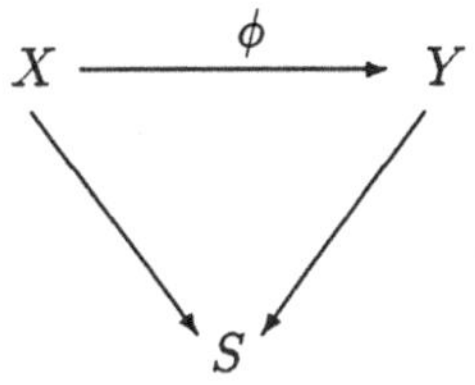

of morphisms of G-schemes, where G acts trivially on S. If $p \in S$ we then have the localized morphism

$$\phi_p : X_p \longrightarrow Y_p$$

of G-S_p-schemes, and the fiber morphism

$$\phi(p) : X(p) \longrightarrow Y(p)$$

of G-$S(p)$-schemes.

(2.4) Theorem. *If for all $p \in S$, $\phi_p : X_p \to Y_p$ is isomorphic to a G-vector bundle over Y_p, then $\phi : X \to Y$ is isomorphic to a G-vector bundle over Y.*

Assume the hypotheses of (2.4). Let V be a fiber of ϕ, which we can take to be a G-module. Then we can think of (2.4) as a reduction of the structure group of the "non-linear bundle" X from the (infinite dimensional) group of G-equivariant *variety automorphisms* of V to the (finite dimensional) group of G-equivariant *linear automorphisms* of V. Even the immediate consequence that $X \to Y$ admits a section is not obvious.

The case $G = \{1\}$ of this theorem is proved in **[BCW]**. The argument there is a somewhat axiomatized rendering of techniques introduce by Quillen in his proof of Serre's Conjecture on projective modules over polynomial rings. The adaptation of those methods to the present equivariant setting involves some technicalities, but no essential difficulties.

It follows easily from Theorem (2.4) that if each $X_p \to Y_p$ is linear or stably linear then $X \to Y$ is, respectively, a vector bundle or stably a vector bundle. Moveover, if each $X_p \to Y_p$ is Asanuma then, under a mild hypothesis on $Y \to S$, $X \to Y$ is Asanuma. This follows from the much stronger theorem that follows.

(2.5) Theorem. *In the setting of* (2.3), *assume that* $Y \to S$ *satisfies "going down"* (2.2), *e.g. that* Y *is faithfully flat over* S. *Then the following conditions are equivalent:*

(a) X *is an Asanuma* G-Y*-scheme.*

(b) X *is flat over* Y *and* $X(p)$ *is an Asanuma* G-$Y(p)$*-scheme for all* $p \in S$.

Taking $Y = S$, we obtain, in view of (1.5):

(2.6) Corollary. *The following conditions are equivalent:*

(a) X *is an Asanuma* G-S*-scheme.*

(b) X *is flat over* S *and* $X(p)$ *is a stably linear* G-$S(p)$*-scheme for all* $p \in S$.

PROOF OF (2.5) (Sketch): That (a) implies (b) is routine. For the converse we are at liberty, thanks to (1.3), to replace X by a G-vector bundle over X. By so doing we can achieve the following condition: There is a G-module V satisfying conditions (i) and (ii) of the next theorem (2.8). Thus (2.5) follows from (2.8).

(2.8) Theorem. *Assume that* X *is flat over* Y. *Let* V *be a* G*-module such that:*

(i) $T_{X/Y} \cong X \times V$ *(as* G*-vector bundles over* X*); and*

(ii) $X(p) \cong Y(p) \times V$ *(as G-Y(p)-schemes) for all* $p \in S$.

Then $X \times V^{m-1} \cong Y \times V^m$ *(as G-Y-schemes), where* $m = 3^{\dim(S)}$. *In particular X is a stably linear, hence Asanuma, G-Y-scheme.*

The proof of (2.8) is by noetherian induction on S. The crucial inductive step is covered by the following "gluing theorem" and its Corollary (2.10), where we use the following notation: Let $\pi \in R$, where $S = \operatorname{Spec}(R)$. For any scheme $Z \to S$ put $Z_\pi = Z \times_S \operatorname{Spec}(R_\pi)$, where $R_\pi = R[\pi^{-1}]$, and $Z' = Z \times_S \operatorname{Spec}(R/\pi R)$. Geometrically, $S' \subset S$ is the divisor defined by vanishing of π, S_π is its complement, and Z', Z_π are the respective inverse images in Z.

(2.9) Theorem. *Let V be a G-module and* $Z = Y \times V$. *Let* $\pi \in R$ *satisfy:*

(0) π *is a non zero-divisor on X and Y.*

(1) $X' \cong Z'$ *(as G-Y'-schemes).*

(2) $X_\pi \cong Z_\pi$ *(as G-*Y_π*-schemes).*

Then there is a G-vector bundle $E \to Z$ *such that* $X \times V \cong E$ *as G-Y-schemes.*

From (2.9) we can easily deduce:

(2.10) Corollary. *In* (2.9), *further assume:*

(3) $T_{X/Y} \cong X \times V$ *(as G-vector bundles over X).*

Then $X \times V^2 \cong Y \times V^3$ *(as G-Y-schemes).*

The proof of (2.9) is based on an elaborate gluing technique, first introduced by Asanuma [**A**] in the non-equivariant case. Using flatness and lifting properties of Asanuma schemes one can strengthen condition (1) to:

$(1)^\wedge$ $\hat{X} \cong \hat{Z}$ (as G-$\hat{Y}$-schemes),

where the " $\wedge$ " denotes π-adic completion. Then from $(1)^\wedge$ and (2) one deduces a description of X as a twisted form $Z(\phi)$ of Z, where $\phi : \hat{Z}_\pi \to Z_\pi$ is a G-Y_π-morphism such that, for some G-Y_π-morphism $\psi : Z_\pi \to Z_\pi$, $\psi \circ \phi$ is the canonical morphism $p : \hat{Z}_\pi \to Z_\pi$. To make $X \cong Z(\phi)$ a vector bundle over Y we would have to "linearize" ϕ, modulo the appropriate equivalence relation. In fact, we are only able to linearize

$$\phi \times p : (Z \times_Y Z)^\wedge_\pi \longrightarrow Z_\pi \times_{Y_\pi} Z_\pi;$$

this is where the need for stabilizing enters. Specifically, one shows that $\phi \times p$ is equivalent to $p \times \phi'$, where ϕ' comes from the differential of ϕ, which is linear over the first Z-factor.

References

[**A**] T. Asanuma, "Polynomial fiber rings of algebras over noetherian rings," Invent. Math. **87** (1987), 101–127.

[**B**] H. Bass, "Linearizing flat families of linear representations," (in preparation).

[**BCW**] H. Bass, E. H. Connell, and D. Wright, "Locally polynomial algebras are symmetric algebras," Invent. Math. **38** (1977), 279–299.

[**BH**] H. Bass and W. Haboush, "Some equivariant K-theory of affine algebraic group actions," Comm. in Algebra **15** (1987), 181–217.

[**M**] H. Matsumura, *Commutative Algebra*, W. A. Benjamin, New York (1970).

Hyman Bass
Department of Mathematics
Columbia University
New York, NY 10027
USA

SPHERICAL VARIETIES
AN INTRODUCTION

MICHEL BRION

When one studies complex algebraic homogeneous spaces it is natural to begin with the ones which are complete (i.e. compact) varieties. They are the "generalized flag manifolds" . Their occurence in many problems of representation theory, algebraic geometry, ... make them an important class of algebraic varieties. In order to study a noncompact homogeneous space G/H, it is equally natural to compactify it, i.e. to embed it (in a G-equivariant way) as a dense open set of a complete G-variety . A general theory of embeddings of homogeneous spaces has been developed by Luna and Vust [**LV**]. It works especially well in the so-called spherical case: G is reductive connected and a Borel subgroup of G has a dense orbit in G/H. (This class includes complete homogeneous spaces as well as algebraic tori and symmetric spaces). A nice feature of a spherical homogeneous space is that any embedding of it (called a spherical variety) contains only finitely many G-orbits, and these are themselves spherical. So we can hope to describe these embeddings by combinatorial invariants, and to study their geometry. I intend to present here some results and questions on the geometry (see [**LV**], [**BLV**], [**BP**], [**Lun**] for a classification of embeddings).

In § 1 I give a local description of a spherical variety, and draw some consequences about its singularities. I also introduce the rank of a spherical homogeneous space G/H; it is the minimal codimension of a K-orbit, where K is a maximal compact subgroup of G. It measures how far G/H is from being complete.

In § 2 I study the topology of a spherical variety, using a cellular decomposition (which generalizes the Bruhat decomposition of flag manifolds). The example of "complete conics" is studied in some detail in 2.4, because of its historical significance.

In § 3 I describe some connections between spherical varieties and moment mappings. They lead to a solution of the following problem: Given a (smooth projective) spherical variety X, of dimension n, describe its second cohomology group $H^2(X, \mathbf{C})$, and compute products of n arbitrary elements of this group. There are interesting applications to classical enumerative problems (see 3.6 and [**DP1**], [**B4**]). As an example, I describe the cohomology algebra of complete conics (3.7).

One of my aims is to call attention to the numerous unsolved problems concerning spherical varieties. The most important one seems to be: to generalize the "Schubert calculus" (which gives a nice picture of the cohomology of a complete homogeneous space). I refer to [**DP3**], [**DGMP**], [**BDP**] for some partial solutions. The reader will find other problems at the end of each section.

Proofs of theorems will be omitted, or only sketched. The result in 3.3 is new, but the proof I know is too long to be published here.

§ 1. Local properties of spherical varieties

1.1. Definitions.

Let G be a connected reductive algebraic group over $\mathbf{C}$. An algebraic subgroup H of G is said to be *spherical* if some Borel subgroup B of G has a dense orbit in the homogeneous space G/H (which is also said to be spherical). We can arrange that the product BH is open in G, and all Borels having this property form a single H-conjugacy class. We call them "opposite Borel subgroups."

Choose an opposite B, and denote by P the set of all $s \in G$ such that $sBH = BH$. Then P is a parabolic subgroup of G which contains B. It will play an important role in the sequel.

1.2. Examples.

(i) G is an algebraic *torus*, and H the trivial group (here any Borel subgroup of G is equal to $G \ldots$).

(ii) G is any connected reductive group, and H is a *parabolic subgroup*. Denote by H^u its unipotent radical. Then a Borel subgroup B is opposite if and only if $B \cap H^u = \{1\}$. Then the parabolic subgroup P is opposite to H in the usual meaning (i.e. $P \cap H$ is maximal reductive in both P and H).

(iii) G is connected reductive. Let σ be an involution of G, and H the subgroup of its fixed points. Then G/H is a *symmetric space*. A Borel subgroup B is opposite exactly when the intersection $B \cap \sigma(B)$ has minimal dimension. P is a minimal σ-split parabolic subgroup, i.e. P and $\sigma(P)$ are opposite, and P is minimal with this property ([**BLV**] 4.1).

There is a classification of *reductive* spherical subgroups H of a reductive group G [**Mik**], [**B2**]. Such "spherical pairs" (G, H) are complexifications of pairs (K, L) where K and L are compact Lie groups and the K-module $L^2(K/L)$ is multiplicity-free [**GS3**].

1.3. The problem of local retractions.

Let G act on an algebraic variety X over $\mathbf{C}$, such that the map $(g, x) \to g.x$ is algebraic (X will be called a *G-variety*). Let $Y = G.x$ be the G-orbit of a given $x \in X$. In general it is difficult to describe the action of G near Y, because (unlike for actions of compact groups) there is no equivariant retraction of a neighbourhood of Y onto Y. Here is an easy example: Let $G = SL_2(\mathbf{C})$ .The linear action of G on the space of quadratic forms in two variables induces an action of G on the complex projective plane X. The conic Y of degenerate quadratic forms, and its complement $X - Y$ are G-orbits. The isotropy group of a point of Y is a Borel subgroup of G, while the isotropy group of a point of $X - Y$ is the normalizer of a torus and is not contained in any Borel. Hence there is no equivariant retraction of a neighbourhood of Y onto Y. However, there is almost a retraction: Let $y \in Y$, let l be the tangent line to Y at y, and let B be the isotropy group of y. Then there is a B-equivariant retraction $r : X - l \to Y - \{y\}$: It associates to $x \in X - l$ the unique $r(x) \in Y - \{y\}$ such that x, y, $r(x)$ are on the same line.

1.4. Action of an algebraic group near a spherical orbit.

Let G, X, Y as in the beginning of 1.3. In view of the preceding example it seems reasonable to try to describe the action of a parabolic subgroup of G on an open set of X which meets Y. When the orbit Y is spherical, we have the following result [**B4**; 1.1].

Theorem. *Let X be a normal G-variety with spherical G-orbit Y. Choose $y \in Y$ and denote by H its isotropy subgroup. Let B, P be as in 1.1, and denote by P^u the unipotent radical of P. Then there exists a Levi subgroup L of P, an open P-stable subset X' of X and a subvariety Z of X' such that*

(i) *Z is affine, L-stable and contains y. Moreover, $Z \cap Y$ is equal to $L.y$, and is closed in Z.*
(ii) *The derived subgroup (L, L) of L fixes y.*
(iii) *The map $(s, z) \in P^u \times Z \to s.z \in X'$ is an isomorphism.*

Furthermore, there exists an affine subvariety Z' of Z, containing Y and stable by the isotropy subgroup L_y, such that $Z' \cap Y = \{y\}$ and such that the map $L \times_{L_y} Z' \to Z$ is an isomorphism.

As a corollary, we obtain the existence of a P-equivariant "retraction" of X' onto $X' \cap Y = P.y$. Moreover, the map $(s, z) \in P^u \times L.y \to s.z \in P.y = B.y$ is an isomorphism.

1.5. Local structure of spherical varieties.

Let us say that the G-variety X is *spherical* if X is normal and some Borel subgroup B of G has a dense orbit. (For $G = T$ a torus this simply means that T has a dense orbit in X. Such varieties, called *toric varieties*, have been thoroughly studied. See [**Oda**] and references therein). Theorem 1.4 can be refined for spherical varieties.

Theorem (see [**B4**], 1.1 and 1.2). *Let X be a spherical variety.*

(i) *X contains only finitely many G-orbits, all of which are spherical. In fact, X contains only finitely many B-orbits [**B1**], [**Vin**]. Let Y, y, H, B and P be as in 1.4. Denote by $X_{Y,B}$ the set of all $x \in X$ such that Y is contained in the closure of the orbit $B.x$.*

(ii) *We can take $X' = X_{Y,B}$ in Theorem 1.4. Moreover, Z (resp. Z') is a spherical L-(resp. L_y-)variety.*

This result means that along a G-orbit a spherical variety looks like an affine spherical variety with a fixed point . As a consequence one can prove that singularities of spherical varieties are "not too bad". Recall that a normal variety V is said to have *rational singularities* if there exists a resolution of singularities $\pi : V' \to V$ such that $R^i\pi_*\mathcal{O}_{V'}$ vanishes for every nonzero i (then this property holds for any resolution).

Corollary ([**B4**] 1.2, [**Pop**] Theorem 10). *The singularities of spherical varieties are rational.*

1.6. Rank of a spherical variety.

Let G, H, B, P, L be as in 1.4. Denote by C the identity component of the center of L. Then the open B-orbit in G/H is isomorphic to $P^u \times (L/L \cap H) = P^u \times (C/C \cap H)$ because L is equal to the product $C.(L, L)$, and (L, L) is contained in $L \cap H$ (Theorem 1.4 (ii)). Define the *rank* of G/H to be the dimension of $C/C \cap H$.

Let us compute the rank for the examples of Part 1.2.

(i) When G is a torus and H the trivial group, then the rank is the dimension of G.

(ii) If H is a parabolic subgroup of G, then the rank of G/H is zero (and this characterizes parabolic subgroups among spherical subgroups).

(iii) If H is the group of fixed points of an involution σ of G, then the rank of G/H has the usual meaning for a symmetric space: it is the dimension of a maximal σ -split torus of G (i.e. a torus A such that $\sigma(a) = a^{-1}$ for every $a \in A$, and maximal with this property).

Define the rank of a spherical variety to be the rank of its dense G-orbit (then one can show that the ranks of the other G-orbits are strictly smaller). By (ii), spherical varieties of rank zero are just complete homogeneous spaces. It can be shown that complete spherical varieties of rank one contain only two or three orbits. By blowing- up the closed orbits one obtains a completion of a homogeneous space by one or two homogeneous divisors. Such objects have been classified by D. N. Ahiezer [**Ahi**].

Let us conclude this section with two open questions on "spherical singularities".

(i) Let X be an affine spherical variety with a fixed point x. Does there exist a $\mathbf{C}^*$ -action on X which commutes with the G-action and has the point $\{x\}$ as unique closed orbit?

(ii) What are the singularities of spherical varieties of rank two ? (see [**B4**] 1.3 for the case of rank one). When G is a torus, one only gets quotients of $\mathbf{C}^2$ by finite cyclic groups ([**Oda**] 1.24).

§ 2. Topology of spherical varieties

2.1. The Bialynicki-Birula decomposition.

Let T be a complex torus acting on a variety X with only finitely many fixed points $x_1, \ldots , x_r$. Then one can choose a one-parameter subgroup $\lambda : \mathbf{C}^* \to T$ such that the fixed points of T and $\mathbf{C}^*$ (acting via λ) are the same; such a λ is said to be "in general position". For $1 \leq i \leq r$ denote by $X(\lambda, x_i)$ the set of all $x \in X$ such that $\lambda(t).x \to x_i$ as $t \to 0$, and call it the *cell* of x_i. If X is projective and smooth, then ([**BB1**], [**BB2**]) each $X(\lambda, x_i)$ is isomorphic to some $\mathbf{C}^{n_i}$ for a non-negative integer n_i. So X has a paving by affine spaces. Moreover, $x_1, \ldots , x_r$ can be ordered so that the closure of each $X(\lambda, x_i)$ is contained in the union of the $X(\lambda, x_j)$'s for $j \geq i$. Hence the cohomology algebra $H^*(X, \mathbf{Z})$ is a free group on the classes of closures $\overline{X(\lambda, x_i)}$. The Poincaré polynomial of X is $\sum_{i=1}^r t^{2n_i}$. If $A^*(X)$ denotes the Chow ring of X (i.e. the group of algebraic cycles on X modulo rational equivalence, the product of two transversal cycles being their intersection), then the natural map $A^*(X) \to H^*(X, \mathbf{Z})$ is an isomorphism.

All these results hold for a projective smooth spherical variety. Indeed, such a variety contains only finitely many G- orbits. Let T be a maximal torus of G. Then an easy lemma ([**DS**] 2.2) shows that T fixes only finitely many points in every homogeneous space of G. So T fixes only finitely many points of X. Proceeding as before we obtain a cellular decomposition of a projective smooth spherical variety, called the *Bialynicki-Birula decomposition*. For a complete homogeneous manifold it coincides with the Bruhat decomposition.

2.2. The Bialynicki-Birula decomposition and B-orbits.

Every spherical variety is a finite union of orbits of a given Borel subgroup of G (Theorem 1.5 (i)). As orbits of a connected solvable algebraic group, they are isomorphic to products of copies of $\mathbf{C}$ and $\mathbf{C}^*$. There is a second coarser decomposition of X into G-orbits, and there is also the Bialynicki-Birula decomposition. A connection between these three decompositions was discovered by DeConcini and Springer [**DS**] for certain compactifications of symmetric spaces , and their results were generalized in [**BL**] using different techniques. In order to state the result we need some more notation.

Choose a Borel subgroup B of G, and a maximal torus T of B. If $\lambda : \mathbf{C}^* \to T$ is a one-parameter subgroup, denote by $P(\lambda)$ the set of all $g \in G$ such that $\lambda(t)g\lambda(t)^{-1}$ has a limit in G as $t \to 0$. It is a parabolic subgroup of G. For a well-chosen λ (inside the "positive Weyl chamber" of T), the group $P(\lambda)$ is equal to B. It follows easily that for every G-variety and every fixed point x of T in X, the cell $X(\lambda, x)$ is B-stable.

We need the following

DEFINITION: A spherical G-variety X with open G-orbit X_G^0 is *toroidal* if the closure of every B-stable divisor in X_G^0 contains no G-orbit.

For example, toric varieties are always toroidal (!). The "complete symmetric varieties" studied by DeConcini, Procesi ... [**DP1**], [**DP2**], [**DS**] can be identified with the toroidal G-varieties whose open G-orbit is a symmetric space.

Theorem ([**BL**] 2.3). *Let X be a toroidal complete G-variety, and let B, T be as before. Then the intersection of any cell $X(\lambda, x)$ with any G-orbit is empty or is a single B-orbit.*

Note that X does not need to be smooth (but it needs to be toroidal; see [**BL**] 2.3). This result shows that B-orbits in X can be described in terms of G-orbits and fixed points of T. Unfortunately, it is in general difficult to compute them (see [**DS**] for complete symmetric varieties).

2.3. A description of the second cohomology group.

We will describe $H^2(X, \mathbf{Z})$ for a smooth complete spherical variety X, in a way which is independant of the Bianynicki- Birula decomposition, and more suited for applications.

Theorem. *Let X be a smooth complete spherical G-variety, and let B be a Borel subgroup of G.*

(i) *The group $H^2(X, \mathbf{Z})$ is generated by the classes of irreducible B-stable divisors of X. The relations are the divisors of the rational functions on X which are eigenvectors of B.*

(ii) *If G has only one closed orbit Y in X, then $H^2(X, \mathbf{Z})$ is freely generated by the classes of irreducible B-stable divisors which do not contain Y.*

REMARK: More generally, if X is any spherical variety (say with only one closed G-orbit Y), then the divisor class group of X is still generated by the classes of irreducible B-stable divisors, with relations as in (i). The Picard group of X is generated by the classes of irreducible B-stable divisors which do not contain Y ([**B4**] §2).

2.4. An example: complete conics.

The set Q consisting of all smooth conics in the complex projective plane $\mathbf{P}^2$ is the homogeneous space $PGL(3)/PO(3)$ where $PGL(3)$ is the automorphism group of $\mathbf{P}^2$, and $PO(3)$ is the projective orthogonal group of a nonsingular quadratic form in three variables. Q is a symmetric space of rank two. Let V be the space of all quadratic forms in three variables, and V^* its dual. Then the projective spaces $\mathbf{P}(V)$ and $\mathbf{P}(V^*)$ are natural compactifications of Q, equivariant with respect to $G = PGL(3)$: We embed Q in $\mathbf{P}(V)$ (resp. $\mathbf{P}(V^*)$) by associating to each smooth conic C its line of equations (resp. the equations of the dual conic C^*, consisting of all tangent lines to C).The closure X of the set of all pairs (C, C^*) in $\mathbf{P}(V) \times \mathbf{P}(V^*)$ is called the space of *complete conics*. It is a (five-dimensional) smooth variety ([**DP1**] Theorem 3.1). We will describe the G- orbits in X, and compute its Poincaré polynomial.

There are three distinct degenerations of a pair (C, C^*), hence three G-orbits in X, outside the open orbit Q:

(i) C degenerates into two distinct lines l and l', and C^* into the set of all lines containing the point $l \cap l'$.

(ii) C^* degenerates into the set of all lines containing one of two fixed points p and p', and C into the double line joining p and p'.

(iii) C degenerates into a double line l, and C^* into the set of all double lines through a given point p on l.

In particular G has only one closed orbit Y in X (the complete conics of type (iii)), isomorphic to the flag variety of $\mathbf{P}^2$.

A maximal torus T of G is the isotropy group of three ordered points x, y, z in $\mathbf{P}^2$, not lying on the same line. It is easy to see that:

(a) T fixes no point of Q.

(b) There are three fixed points of T in the G-orbit of complete conics of type (i) (resp. (ii)), obtained by taking two sides of the triangle xyz as l and l' (resp. two vertices of this triangle as p and p').

(c) There are six fixed points of T in the complete conics of type (iii), i.e. the six flags defined by the triangle xyz.

Hence T fixes 12 points of X: the rank of $H^*(X,\mathbf{Z})$ is 12.

Let B be a Borel subgroup of G. Then B is the isotropy group of a flag (p,l) in $\mathbf{P}^2$. A B-orbit in Q is described by the relative position of a flag and a conic. Hence there are two B-orbits of codimension one in Q: the set of conics containing p (resp. tangent to l). From Theorem 2.3 (ii) we deduce that their closures (classically denoted by μ and ν) form a basis of $H^2(X,\mathbf{Z})$. In particular, its rank is equal to 2. From the fact that $H^*(X,\mathbf{Z})$ is of rank 12, and Poincaré duality, we conclude that *the Poincaré polynomial of X is* $1+2t^2+3t^4+3t^6+2t^8+t^{10}$. The algebra $H^*(X,\mathbf{Z})$ will be described by generators and relations in 3.7 below.

§ 3. Spherical varieties and the moment map

3.1. The moment map.

Let V be a representation space of a connected reductive complex group G. Choose a maximal compact subgroup K of G, and a K-invariant positive definite hermitian form on V; we denote by a dot the associated scalar product. One can then canonically define a K-invariant Kähler structure on the projective space $\mathbf{P}(V)$, which is, in particular, a symplectic variety (the symplectic form ω being the imaginary part of the Kähler form). Define a map μ from X to $\mathcal{K}^*$ (the dual of the Lie algebra $\mathcal{K}$ of K) by $\langle\mu(x),A\rangle = (\tilde{x}.A\tilde{x})(\tilde{x}.\tilde{x})^{-1}$ where $x\in\mathbf{P}(V)$; $\tilde{x}$ is a representant of x in V; $A\in\mathcal{K}$. It is easy to see that μ is K-equivariant and that its differential $d\mu$ verifies: $\langle d\mu_x(\xi),A\rangle = \omega_x(\xi,A_x)$ for every $x\in X$, $\xi\in T_x\mathbf{P}(V)$, $A\in\mathcal{K}$. This means that μ is a *moment map* for the symplectic action of K on $\mathbf{P}(V)$ (**[GS1&2]**, **[Kir]**, **[Nes]**).

If X is any closed smooth algebraic subvariety of $\mathbf{P}(V)$ which is G-stable, then X inherits a K-invariant Kähler structure, and the restriction of μ to X is still a moment map for the induced symplectic structure on X.

The image $\mu(X)$ has a nice convexity property. As $\mu(X)$ is a K-stable subset of $\mathcal{K}^*$, it is described by its intersection with a fundamental domain C of K acting on $\mathcal{K}^*$. We can choose C to be a Weyl chamber in the dual $\mathcal{T}^*$ of a Cartan algebra $\mathcal{T}$ of $\mathcal{K}$.

Theorem ([**GS1&2**], [**Kir**], [**Nes**]). *The intersection $\mu(X) \cap C$ is a convex polyhedron with rational vertices.*

(Recall that $\mathcal{T}^*$ contains the lattice $\mathcal{T}^*_{\mathbf{Z}}$ of weights of the maximal torus T of K with Lie algebra $\mathcal{T}$. Hence we can speak of integral and rational points of $\mathcal{T}^*$).

3.2. A characterization of spherical varieties.

Keep the same notation as above. Let $\pi : \mathcal{K}^* \to \mathcal{K}^*/K$ be the quotient by K. The restriction of π to C is an isomorphism. The composition $\overline{\mu} = \pi \circ \mu$ is a K-invariant map from X to a linear space.

Theorem ([**B3**] 5.1). *Let X be a G-stable closed smooth subvariety of $\mathbf{P}(V)$. Then the following conditions are equivalent:*

(i) *X is spherical*
(ii) *The fibers of $\overline{\mu}$ are K-orbits.*

If either (i) or (ii) is verified, then the rank of X is the dimension of $\overline{\mu}(X)$, i.e. of $\mu(X) \cap C$. It is also the minimal codimension of the K-orbits in X.

For a smooth projective spherical G-variety X, we have thus a realization of the quotient by K as a map to a convex polyhedron (because we can identify $\overline{\mu}(X)$ and $\mu(X) \cap C$). This raises a number of questions: is this map a quotient in the C^∞ category ? Is it possible to obtain topological information about X from the contractibility of X/K ?

3.3. An integral formula.

Consider as before a closed G-subvariety X of $\mathbf{P}(V)$. Let Γ be the intersection of $\mu(X)$ with the Weyl chamber $C \subset \mathcal{T}^*$. Denote by R the root system of G (with respect to the maximal torus associated to $\mathcal{T}$), R^+ the set of positive roots defined by the choice of C, and E the set of all positive roots which are orthogonal to Γ. Let W be the Weyl group and W_E the subgroup which stabilizes E.

The symplectic form ω on X gives us a measure

$$\frac{\omega^n}{n!}$$

called the *Liouville measure* (n is the complex dimension of X). Let us determine the push-forward of this measure by the moment map, by computing its Fourier transform:

$$A \in \mathcal{K} \to \int_X exp\langle \mu(x), A \rangle \frac{\omega_x^n}{n!}$$

By K-equivariance of μ we may assume that $A \in \mathcal{T}$.

Theorem. *Let $X \subset \mathbf{P}(V)$ be a spherical variety. Then for every $A \in \mathcal{T}$ we have*

$$\int_X exp\langle \mu(x), A \rangle \frac{\omega_x^n}{n!}$$
$$= \prod_{\alpha \in R+} \alpha(A)^{-1}. \sum_{w \in W/W_E} det(w) \prod_{\alpha \in E} (w\alpha)(A) \int_\Gamma exp\langle w\gamma, A \rangle d\gamma$$

where $d\gamma$ is the Lebesgue measure on $\mathcal{T}^*$, normalized so that the torus $\mathcal{T}^*/\mathcal{T}^*_{\mathbf{Z}}$ has volume one.

REMARK: This result can be generalized to G–varieties which are not necessarily spherical. The above formula holds providing that the Lebesgue measure $d\gamma$ is replaced by $M(\gamma)d\gamma$, where the function M is defined as follows : For every $\gamma \in \Gamma$, the isotropy group K_γ acts on the fiber $\mu^{-1}(\gamma)$ and (for a general γ) the quotient $\mu^{-1}(\gamma)/K_\gamma$ has a canonical symplectic structure ([**DH1**] §2). Define $M(\gamma)$ to be its volume, with respect to its Liouville measure. It can be shown that the function M is locally polynomial on Γ (for a spherical variety we know from 3.2 that $\mu^{-1}(\gamma)$ is a single K_γ–orbit so that M is constant, equal to one). Full details and proofs will appear elsewhere. The case when G is a torus has been studied by Duistermaat and Heckman ([**DH1**], [**DH2**]).

Note that for a toric variety the above formula becomes very simple. In fact we have the following

Corollary. *Let $X \subset \mathbf{P}(V)$ be a closed G-variety, where G is a torus acting linearly on V. Let Γ be the image of the moment map. If G has a dense orbit in X, then the push-forward of the Liouville measure on X is the Lebesgue measure on Γ.*

PROOF: By the theorem we have

$$\int_X exp\langle \mu(x), A \rangle \frac{\omega_x^n}{n!} = \int_\Gamma exp\langle \gamma, A \rangle d\gamma$$

This fact has been noticed by Atiyah [**Ati**].

3.4. The degree of a spherical variety in a projective space.

Recall that the *degree* d of a closed n-dimensional subvariety X of $\mathbf{P}(V)$ is the number of points of intersection of X and a general subspace of codimension n (such a space is transversal to X). As the symplectic form ω and a general hyperplane section of X have the same cohomology class, we have $d = \int_X \omega^n$. Notation being as before, we can state the following

Theorem ([**B4**] 4.1). *The degree of* X *is equal to*

$$n! \int_\Gamma \prod_{\alpha\in R^+ - E} \frac{(\gamma,\alpha)}{(\rho,\alpha)} d\gamma$$

where ρ *is half the sum of the positive roots.*

PROOF: Replace A by tA in Theorem 3.3 and let $t \to 0$. The left-hand side has a limit, which is the volume of X, i.e. $\frac{d}{n!}$. Expand the right-hand side in a series of powers of t. The constant term is

$$\prod_{\alpha\in R^+} \alpha(A)^{-1}. \sum_{w\in W/W_E} \prod_{\alpha\in A} (w\alpha)(A) \int_\Gamma \frac{\langle w\gamma, A\rangle^{N-N_E}}{(N-N_E)!} d\gamma$$

where N (resp.N_E) is the cardinality of R^+ (resp. E). Now use the following identity on root systems ([**B4**] Lemme 4.2.1):

$$\sum_{w\in W/W_E} det(w).(\prod_{\alpha\in E} w\alpha).(w\gamma)^{N-N_E}$$

$$= (N-N_E)! \prod_{\alpha\in R^+} \alpha. \prod_{\alpha\in R^+ - E} \frac{(\gamma,\alpha)}{(\rho,\alpha)}.$$

3.5. Common zeroes of Laurent polynomials.

As a corollary of all this theory, let us prove a result of A.Kushnirenko [**Kus**]. He considers systems of equations $\sum a_m x^m = 0$ where $m = (m_1, \ldots, m_n)$ is a multi-index of (positive or negative) integers, and x^m is the monomial $x_1^{m_1} \cdots x_n^{m_n}$. We look for their solutions in $(\mathbf{C}^*)^n$.

Theorem ([**Kus**] Théorème III'). *Consider a system S:* $f_1 = \cdots f_n = 0$ *where the* f_i*'s are Laurent polynomials. Let* Γ *be the convex hull in* $\mathbf{C}^n$ *of all the* m*'s such that* x^m *appears in some* f_i*. If the* f_i*'s are sufficiently* *general, then the number of solutions of S in* $(\mathbf{C}^*)^n$ *is equal to n! times the volume of* Γ.

SKETCH OF PROOF (see also [**Ati**] §3 and §5): Let $m_1, \cdots m_p$ be the integral points in Γ. Define a vector space V with basis $\{e_1, \ldots, e_p\}$, and let $T = (\mathbf{C}^*)^n$ act on V by: $x.e_i = x^{m_i} e_i$. Let X denote the closure in $\mathbf{P}(V)$ of the T-orbit of $x_0 = [e_1 + \cdots + e_p]$. As each f_i only involves monomials x^{m_j}, one can interpret each equation $f_i(x) = 0$ as a hyperplane section of $T.x_0$. If the f_i's are sufficiently general, they are transversal to the boundary $X - T.x_0$, and so they cut $T.x_0$ in d points, where d is the degree of X in $\mathbf{P}(V)$. It is easily seen that the image $\mu(X)$ is equal to Γ. Moreover, Theorem 3.4 applied to X gives: $d = n! \int_\Gamma d\gamma$.

3.6. Some problems in enumerative geometry.

In the preceding example we have in fact solved the following problem: Given n sufficiently general hypersurfaces on an algebraic torus $(\mathbf{C}^*)^n$, compute the cardinality of their intersection. This kind of problem was raised by many geometers in the nineteenth century, but instead of a torus they considered spaces which arise naturally in classical projective geometry: grassmanians, flag varieties, quadrics, cubic curves ... It turns out that these spaces are very often homogeneous, and that some of them are spherical. In the latter case, we can solve the problem of "intersecting n divisors" as follows [**DP2**], [**B4**].

Let $D_1, \ldots, D_n$ be hypersurfaces in an n-dimensional spherical homogeneous space G/H. From [**DP2**] we know that there exists a smooth compactification X of G/H such that the closures $\overline{D_1}, \ldots, \overline{D_n}$ contain no G–orbit of X. As X has only finitely G–orbits, a simple generalization of a transversality theorem of Kleiman ([**DP1**] 6.1) shows that, for generic $g_i \in G$, the translates $g_1\overline{D_1}, \ldots, g_n\overline{D_n}$ are transversal (moving the $\overline{D_i}$'s by $g_i \in G$ means putting them in general position). Moreover, the cardinality of the intersection of the $g_i D_i$'s is the product $[\overline{D_1}] \cdots [\overline{D_n}]$ of their cohomology classes in $H^*(X, \mathbf{Z})$.

So we are reduced to describing $H^2(X, \mathbf{Z})$ (which has been done in 2.3) and finding the product of n arbitrary elements of this group. It is enough to compute the powers $c_1(L)^n$, where $c_1(L)$ is the Chern class of a *positive* line bundle L (because we can find positive $L_1, \ldots, L_r$ such that the $c_i(L)$'s generate the vector space $H^2(X, \mathbf{Q})$; then for $x_i > 0$ the line bundle $L = L_1^{x_1} \cdots L_r^{x_r}$ is positive; by expanding $c_1(L)^n$ we find every monomial in the L_i's). But such an L defines an embedding of X into a projective space $\mathbf{P}(V)$, and $c_1(L)^n$ is just the degree of X in $\mathbf{P}(V)$, which is given by the integral formula 3.4.

We sketch how this process works for a (very standard) example: Compute the number N of conics which are tangent to 5 given conics in general position [**Kle**]. Denote by D the hypersurface of $G/H = PGL(3)/PO(3)$ consisting of all conics which are tangent to a given conic. The closure of D in the variety X of complete conics (see 2.4) does not contain the closed G-orbit Y. The class of $\overline{D}$ in $H^2(X, \mathbf{Z})$ is equal to $2(\mu+\nu)$ where μ, ν are defined in 2.4. So $N = 2^5(\mu+\nu)^5$. Now X is a subvariety of $\mathbf{P}(V) \times \mathbf{P}(V^*) \subset \mathbf{P}(V \otimes V^*)$, and μ (resp. ν) is the pullback of the hyperplane section of $\mathbf{P}(V)$ (resp. $\mathbf{P}(V^*)$). So $\mu+\nu$ is the hyperplane section of X in $\mathbf{P}(V \otimes V^*)$, and $N = 32d$ where d is the degree of X in $\mathbf{P}(V \otimes V^*)$. Consider the following sketch of $\Gamma \subset C \subset \mathcal{T}^*$ (see 3.1 to 3.3).

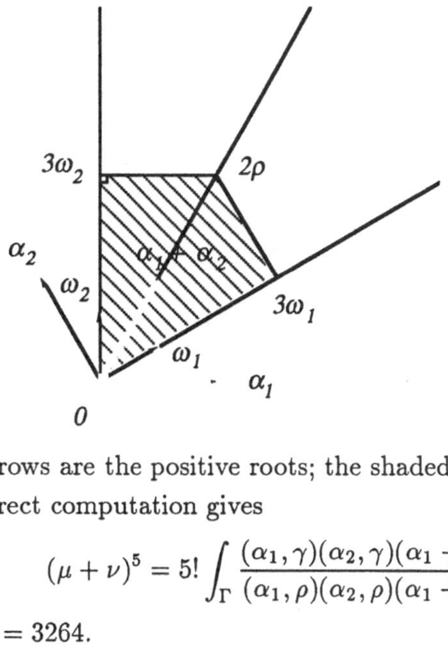

The arrows are the positive roots; the shaded region is Γ.

A direct computation gives

$$(\mu+\nu)^5 = 5! \int_\Gamma \frac{(\alpha_1, \gamma)(\alpha_2, \gamma)(\alpha_1+\alpha_2, \gamma)}{(\alpha_1, \rho)(\alpha_2, \rho)(\alpha_1+\alpha_2, \rho)} d\gamma = 102$$

and $N = 3264$.

3.7. Cohomology of complete conics.

We continue to use the notation of 2.4. Using the above machinery we now describe the cohomology algebra (over $\mathbf{Q}$) of the space X of complete conics.

Theorem. *The algebra $H^*(X, \mathbf{Q})$ is generated by μ and ν; the ideal of relations is generated by $2\mu^3 - 3\mu^2\nu + 3\mu\nu^2 - 2\nu^3$ and $4\mu^4 - 3\mu^2\nu^2 + 4\nu^4$.*

PROOF: Let us compute all monomials of degree 5 in μ and ν (they are called *characteristic numbers* of complete conics; see [**Kle**]). Let m, n be positive integers. Then $(m\mu + n\nu)^5$ is the degree of X in the projective space $\mathbf{P}(V^{\otimes m} \otimes (V^*)^{\otimes n})$. The associated polygon Γ has 0, $(2m+n)\omega_1$, $(m+2n)\omega_2$, $2(m\omega_1 + n\omega_2)$ as vertices, where ω_1, ω_2 are the fundamental weights of $PGL(3)$ ([**B4**] 2.7). We find that $(m\mu+n\nu)^5 = m^5 + 10m^4n + 40m^3n^2 + 40m^2n^3 + 10mn^4 + n^5$, i.e. $\mu^5 = \nu^5 = 1$, $\mu^4\nu = \mu\nu^4 = 2$, $\mu^3\nu^2 = \mu^2\nu^3 = 4$. These equalities can also be derived directly ([**Kle**] p.12).

Now let us prove that $\{\mu^2, \mu\nu, \nu^2\}$ is a basis of $H^4(X, \mathbf{Q})$. If $x\mu^2 + y\mu\nu + z\nu^2 = 0$ is a linear relation, then multiplying successively by μ^3, $\mu^2\nu$, $\mu\nu^2$, ν^3 we obtain $x+2y+4z = 0$, $2x+4y+4z = 0$, $4x+4y+2z = 0$, $4x+2y+z = 0$ which implies that $x = y = z = 0$.

We can prove in the same way that $\{\mu^3, \mu^2\nu, \mu\nu^2, \nu^3\}$ is a basis of $H^6(X, \mathbf{Q})$, and that $\{\mu^4, \nu^4\}$ is a basis of $H^8(X, \mathbf{Q})$. So μ,ν generate $H^*(X, \mathbf{Q})$. It is also easily verified that $2\mu^3 - 3\mu^2\nu + 3\mu\nu^2 - 2\nu^3$ has a zero product with μ^2, $\mu\nu$ and ν^2. By Poincaré duality it must be zero. Similarly $4\mu^4 - 3\mu^2\nu^2 + 4\nu^4 = 0$.

Consider now the graded commutative algebra A over $\mathbf{Q}$ generated by x and y of degree 2, with relations $2x^3 - 3x^2y + 3xy^2 - y^3$ and $4x^4 - 3x^2y^2 + 4y^4$. As the two relations have no common divisor, A is a complete intersection, so that its Poincaré polynomial is

$$\begin{aligned}\frac{(1-z^6)(1-z^8)}{(1-z^2)^2} &= (1+z^2+z^4)(1+z^2+z^4+z^6)\\ &= 1 + 2z^2 + 3z^4 + 3z^6 + 2z^8 + z^{10}\end{aligned}$$

Obviously we have a surjective morphism from A to $H^*(X, \mathbf{Q})$. Since both sides have the same Poincaré polynomial, they are isomorphic.

Let me conclude with two *problems* .

(i) *How does one describe the cohomology algebra of a (smooth projective) spherical variety ?*

(ii) *How can one generalize our methods to determine the second cohomology group and characteristic numbers of a non-spherical variety ?*

References

[**Ahi**] D. N. Ahiezer, *Equivariant completion of homogeneous algebraic varieties by homogeneous divisors*, Ann. Glob. Analysis and Geometry **1** (1983), 49–78.

[**Ati**] M. F. Atiyah, *Angular momentum, convex polyhedra and algebraic geometry*, Proceedings of the Edinburgh Math. Soc. **26** (1983), 121–138.

[**BB1**] A. Bialynicki-Birula, *Some theorems on actions of algebraic groups*, Ann. of Math. **98** (1973), 480–497.

[BB2] A. Bialynicki-Birula, *Some properties of the decomposition of algebraic varieties determined by actions of a torus*, Bull. Acad. Polo. Ser. Sci. Math. Astronom. Phys. **24** (1976), 667–674.

[BDP] E. Bifet, C. DeConcini, C. Procesi, *Cohomology of complete symmetric varieties*, Manuscript 1988.

[B1] M. Brion, *Quelques propriétés des espaces homogènes sphériques*, manuscripta math. **55** (1986), 191–198.

[B2] M. Brion, *Classification des espaces homogènes sphériques*, Compositio Math. **63** (1987), 189–208.

[B3] M. Brion, *Sur l'image de l'application moment*, Séminaire d'algèbre, Springer Lecture Notes 1296.

[B4] M. Brion, *Groupe de Picard et nombres caractéristiques des variétés sphériques*, To appear in the Duke Mathematical Journal.

[BL] M. Brion, D. Luna, *Sur la structure locale des variétés sphériques*, Bull. Soc. Math. France **115** (1987), 211–226.

[BLV] M. Brion, D. Luna, T. Vust, *Espaces homogènes sphériques*, Invent. math. **84** (1986), 617–632.

[BP] M. Brion, F. Pauer, *Valuations des espaces homogènes sphériques*, Comment. Math. Helv. **62** (1987), 265–285.

[DH1] J. J. Duistermaat, G. Heckman, *On the variation in the cohomology of the symplectic form of the reduced phase space*, Invent. math. **69** (1982), 259–268.

[DH2] J. J. Duistermaat, G. Heckman, *Addendum to "On the variation ... "*, Invent. math. **72** (1983), 153–158.

[DP1] C. DeConcini, C. Procesi, *Complete symmetric varieties*, Invariant theory, Springer Lecture Notes 996.

[DP2] C. DeConcini, C. Procesi, *Complete symmetric varieties II*, Algebraic groups and related topics, North-Holland 1985.

[DP3] C. DeConcini, C. Procesi, *Cohomology of compactifications of algebraic groups*, Duke Math. J. **53** (1986), 585–594.

[DGMP] C. DeConcini, M. Goresky, R. MacPherson, C. Procesi, *On the geometry of complete quadrics*, Comment. Math. Helv. **63** (1988), 337–413.

[DS] C. DeConcini, T. A. Springer, *Betti numbers of complete symmetric varieties*, Geometry Today (Birkhäuser 1985).

[GS1&2] V. Guillemin, S. Sternberg, *Convexity properties of the moment mapping I* & II, Invent. math. **67** (1982), 491–513; Invent. math. **77** (1984), 533–546.

[GS3] V. Guillemin, *Multiplicity-free spaces*, J. Diff. Geometry **19** (1984), 31–56.

[Kir] F. Kirwan, *Convexity properties of the moment mapping III*, Invent. math. **77** (1984), 547–552.

[Kle] S. L. Kleiman, *Chasles' enumerative theory of conics: A historical introduction*, Aarhus Universitet, Preprint.

[Kus] A. G. Kushnirenko, *Polyèdres de Newton et nombres de Milnor*, Invent. math. **32** (1976), 1–31.

[**LV**] D. Luna, T. Vust, *Plongements d'espaces homogènes*, Comment. Math. Helv. **58** (1983), 186–245.
[**Lun**] D. Luna, *Report on spherical varieties*, Manuscript 1986.
[**Mik**] I. V. Mikityuk, *On the integrability on invariant hamiltonian systems with homogeneous configuration spaces*, Math. Sbornik **129** [**171**] (1986), 514–534.
[**Nes**] L. Ness, *A stratification of the nullcone via the moment map*, Amer. J. Math. **106** (1984), 1281–1330.
[**Oda**] T. Oda, *Convex bodies and algebraic geometry (An introduction to the theory of toric varieties)*, (Springer-Verlag), Ergebnisse der Mathematik **15**.
[**Pop**] V. L. Popov, *Contraction of the actions of reductive algebraic groups*, Math. of the USSR Sbornik **58** (1987), 311–335.
[**Vin**] E. B. Vinberg, *Complexity of the actions of reductive groups*, Funct. Anal. and its Appl. **20** (1986), 1–13.

Michel Brion
Institut Fourier
Université de Grenoble
F-38402 Saint-Martin-d'Hères
FRANCE

HOMOLOGY PLANES
AN ANNOUNCEMENT AND SURVEY

TAMMO TOM DIECK AND TED PETRIE

We review recent advances dealing with homology planes, i.e., *non singular affine acyclic surfaces* over the complex numbers **C**. Here acyclic means vanishing integral reduced homology. From a topological point of view these are the simplest affine surfaces.

Before mentioning our results, we give some of the history of this topic. The first example of a homology plane different from the standard complex plane $\mathbf{C}^2$ is due to Ramanujam [**R**]. He exhibited a homology plane R which is even contractible. This produced a counterexample to the existing conjecture that such a surface was $\mathbf{C}^2$. He went on to characterize $\mathbf{C}^2$ as the only homology plane which is simply connected at infinity. Ramanujam's paper appeared in 1972.

It wasn't until 1987 that other homology planes were discovered. In that year Gurjar-Miyanishi [**GM**] produced an infinite number of examples. All but one of these was of (logarithmic) Kodaira dimension one. One was a new homology plane of Kodaira dimension 2. (The Ramanujam example has Kodaira dimension 2 as well.) Their paper in fact produced all homology planes of Kodaira dimension 1, but classification remains open because there are redundancies in their list of these surfaces. In [**tDP**] the authors use the work of [**GM**] to classify the contractible homology planes of Kodaira dimension 1 and to show that a large class of these actually occur as hypersurfaces in $\mathbf{C}^3$.

Theorem [**tDP**]. *Infinitely many non-isomorphic homology planes of Kodaira dimension 1 occur as hypersurfaces in* $\mathbf{C}^3$.

(This is made more precise in loc.cit.)

It wasn't previously known that homology planes could occur as hypersurfaces, but it would be surprising to the authors if all homology planes were hypersurfaces.

In their paper [**GM**], Gurjar-Miyanishi ask whether there are an infinite number of homology planes of Kodaira dimension 2. We announced an affirmative answer at the Rutgers Conference (see Theorem F below). Theorems D, E and F show that there are several countably infinite families of homology planes of Kodaira dimension 2; these families arise in a way to be described from a configuration of lines in the projective plane $\mathbf{P}^2$. At the same time Miyanishi had produced an infinite family of homology planes of Kodaira dimension 2 (see [**M**]). While it is not obvious, Miyanishi's family occurs as one of the families in our list. That this is the case is a consequence of the theory of Cremona transformations briefly discussed in section 3. In response to queries from us, Neumann also independently produced an infinite family of homology planes of Kodaira dimension 2 using the same method as [**GM**]. These examples also appear as one family in our list.

Homology planes are also interesting from the point of view of algebraic transformation groups. The basic homology plane is $\mathbf{C}^2$. Its full automorphism group is known. It is the amalgamated product of the group of affine automorphisms of the complex plane and the Jonquière group of automorphisms of the complex plane (see [**K**] in this volume). From this it follows that every algebraic action of a reductive group on $\mathbf{C}^2$ is isomorphic to a linear representation.

It is known that the Kodaira dimension of a variety influences its automorphism group. In particular, a surface of Kodaira dimension 2 has finite automorphism group (see [**I**]). This contrasts with the automorphism group of $\mathbf{C}^2$, as seen above. The Kodaira dimension of $\mathbf{C}^2$ is $-\infty$ and it is the only homology plane of Kodaira dimension $-\infty$. It is part of the Homology Plane Conjecture [**P**] that the automorphism group of a homology plane of Kodaira dimension 2 is trivial. This conjecture asserts that the only homology plane which admits a non trivial finite order automorphism, is the standard complex plane. The conjecture holds for the homology planes of Kodaira dimensions 0 and 1 and has been verified by the authors for many of the planes of Kodaira dimension 2. It was this conjecture as part of the larger area of transformation groups on acyclic varieties which motivated the study here of homology planes.[1]

To indicate the subtlety of the conjecture, we remark that it is false in the real algebraic setting. Indeed, the contractible homology planes of Kodaira dimension 1 are all hypersurfaces as mentioned above. They are realized by polynomials with real coefficients and hence they define varieties invariant under complex conjugation.

We now want to present some of the ideas, tools and invariants which

[1] **Added in Proof:** The first author has recently produced a counterexample to the Homology Plane Conjecture.

play a role in the construction and identification of homology planes. For this we need to introduce some notation.

By far the most important geometric construction from surface theory that we need is the operation of *blowing up* a smooth point p of a surface W. This produces a new surface X and a map $c : X \to W$ called *the blow-up map* or *contraction*. The inverse image E of p is called the *exceptional fiber*. Then c maps $X \setminus E$ isomorphically onto $W \setminus \{p\}$. This construction may be iterated by blowing up a point of X, etc. This produces a series of surfaces and contractions to their predecessors and so a map from the final surface to the initial surface which is a composition of individual contractions. Such a map will also be called a contraction. If $c : X \to W$ is a contraction, we call

$$\Sigma = \{x \in W \mid c^{-1}(x) \text{ is not one point}\}$$

the *exceptional set* for c. The *exceptional divisors* (also called exceptional fibers) of the contraction c are by definition the irreducible components of $c^{-1}(x)$ for $x \in \Sigma$. We denote these by $\{E_u \mid u \in U\}$. If Z is a curve in W, its proper transform (under c) is the curve $\tilde{Z} = \overline{c^{-1}(Z \setminus \Sigma)}$ in X. Here $\overline{A}$ denotes the Zariski closure of A.

Let V be an affine surface. If $V = X \setminus Y$, where X is a projective surface and $Y \subset X$ a curve (which might be reducible), and if there is a contraction $c : X \to \mathbf{P}^2$, then we call $D = c(Y)$ a *plane divisor* of V.

Proposition A. *If V is rational, then V has a plane divisor.*

If $V = X \setminus Y$ and X has a contraction to $\mathbf{P}^2$, this is obvious. In general there is an X' and Y' such that $V = X' \setminus Y'$ and X' contracts to $\mathbf{P}^2$.

The fact that a homology plane has at least one plane divisor is Corollary B below. Plane divisors, when they exist for an affine variety are not unique. It would be helpful to have a "simple" plane divisor for each homology plane. All known homology planes have a *linear* plane divisor, i.e., one whose irreducible components are lines. Use of Cremona transformations is central in this verification.

Theorem [GS]. *Homology planes are rational.*

Corollary B. *Homology planes have plane divisors.*

To motivate this terminology, it is instructive to present Ramanujam's construction of his homology plane. Start with a cubic with a cusp and a regular conic in the projective plane which intersect in two regular points with intersection multiplicity 1 and 5. Blow up the point of multiplicity

1. This produces a projective surface X and a contraction $c : X \to \mathbf{P}^2$. Let Y be the proper transform of the union of the cubic and the conic, i.e., Y is the closure of the inverse image of the cubic and conic with the exceptional fiber deleted. Then $V = X \setminus Y$ is the Ramanujam surface which, as mentioned, is a homology plane. Its plane divisor is the union of the cubic and conic. Using the theory of Cremona transformations one can show that there is a configuration of lines in $\mathbf{P}^2$ which is also a plane divisor of V, i.e., there is another projective surface X', a curve $Y' \subset X'$ and a contraction $c' : X' \to \mathbf{P}^2$ such that $V = X' \setminus Y'$ and that $c'(Y')$ is a configuration of lines.

As mentioned we would like to find a "simple" plane divisor for each homology plane. One measure of simplicity is the nature of singularities. We say that a curve in a non-singular surface has *tidy* singularities if the branches of the curve through each of its points intersect transversally. A plane divisor is tidy if it has only tidy singularities.

Theorem C. *Each homology plane has a tidy plane divisor.*

This is proved in Corollary 3.6. It uses Cremona transformations and a Theorem of Max Noether.

A *line configuration* is a union of lines in $\mathbf{P}^2$. The word configuration refers to a union of irreducible curves in any non-singular surface. We abuse notion by using the word configuration to also mean the curve whose irreducible components are the members of the configuration. A configuration is tidy if viewed as a curve, it has only tidy singularities. A line configuration is automatically tidy.

There are many examples where a homology plane is presented with a plane divisor which is not linear, but a Cremona transformation may be used to construct a new plane divisor which is linear. This technique is discussed in section 3.

The above discussion leads to these

Basic Questions.

(i) *Which tidy configurations can occur as plane divisors of homology planes?*

(ii) *Does every homology plane have a linear plane divisor?*

(iii) *Which configurations of lines can occur as plane divisors of homology planes?*

(iv) *How does one describe the set of homology planes which have linear plane divisors?*

The first two questions are open and are tantamount to classifying homology planes. Theorem D answers (iii) and Theorem E answers (iv).

Theorem D. *Every homology plane which has a linear plane divisor has one in the following list:*

(i) *L(1,n+1): This configuration consists of $n+1$ lines where n pass through one point while the remaining line does not pass through this point.*

(ii) *L(2): This consists of 4 lines in general position.*

(iii) *L(3): This consists of 5 lines where 4 are in general position and the fifth is a line through two points of intersection in the 4 line configuration.*

(iv) *L(4): This consists of 7 lines with 3 points of valence 2 and 6 points of valence 3. (The valence of a point is the number of lines through the point.)*

(v) *L(5): This is the first of two configurations of 9 lines with 6 points of valence 2, 6 points of valence 3 and 2 points of valence 4.*

(vi) *L(6): This is the second of two configurations of 9 lines and the preceding valence properties.*

(vii) *L(7): This is a configuration of 10 lines. There are 8 points of valence 2, 7 of valence 3, 1 of valence 4 and 1 of valence 5.*

One important invariant of a curve D in a projective surface W is its dual graph $T = T(W, D)$ (see section 2). When T is a minimal tree, we call $T = T(W, D)$ the *compactification tree* for the surface $W \setminus D$. In general this is not a well defined function of $W \setminus D$ as there are other pairs consisting of a projective surface and curve on it whose complement is $W \setminus D$. None the less, in the examples of this report it is unique and we proceed as if this is always the case.

The compactification tree of a homology plane is an important invariant. E.g., it plays a central role in studying the automorphisms of the homology plane (see [**P**]). It determines the fundamental group at infinity of the variety. (See e.g. [**GS**].) Using 2.7(iv), one determines the compactifaction trees of the homology planes constructed in 2.11. The procedures in section 2 give the method of computing the compactification trees of all homology planes having a linear plane divisor. These trees all arise from dual graphs of the line configuration to be realized as plane divisors and among other things are involved with certain continued fraction expansions. It turns out that the compactification trees of the homology planes which have a linear plane divisor can therefore be more efficiently described in terms of an associated special compactification tree which incorporates the continued fraction information in a different way. This is described in section 2. For example, the compactification trees of the homology planes having a plane divisor consisting of 4 lines in general position in $\mathbf{P}^2$ can have an arbitrarily large number of vertices while the associated special compactification trees all have 4 vertices.

We now briefly describe all the homology planes which have some linear plane divisor. We can suppose that the plane divisor is in the list given in Theorem D.

Theorem E. *Let V be a homology plane which has a linear plane divisor. Then V has a linear plane divisor L in the list in Theorem D and V arises from L via the following procedure:*

(i) STEP 1: *Blow up all points of valence greater than 2, i.e., those points which lie on at least three lines in the line configuration L. This produces a surface X' and and contraction $\pi' : X' \to \mathbf{P}^2$.*

(ii) STEP 2: *Form a new configuration Y' of curves in the surface X. This new configuration consists of the proper transforms of the lines in the original configuration together with a selection of some of the exceptional fibers produced in the preceding step. Each possible selection of exceptional fibers together with the proper transforms of the lines in $\mathbf{P}^2$ gives a different configuration which must be considered. Note that $Y' = \overline{\pi'^{-1}(L) \setminus F}$ where F is the union of the exceptional fibers of π' not selected.*

(iii) STEP 3: *Choose vertex and edge functions for the dual graph T of the new configuration. (See section 2 for definitions.)*

(iv) STEP 4: *Blow up edges and vertices of T according to the edge and vertex functions and realize these combinatorial blow-ups geometrically. (See section 2.) (These combinatorial operations on T produce a new tree which is the dual graph of a curve on a surface obtained from the surface in the preceding step by repeated blow-ups mimicking the combinatorial blow-ups on T.) This produces a surface X and a contraction $\pi : X \to X'$.*

(v) STEP 5: *Let $Y \subset X$ be defined by $Y = \overline{\pi^{-1}(Y') \setminus F'}$ where F' is the union of the exceptional divisors of π of weight -1.*

(vi) STEP 6: *Compute $\mathrm{Pic}(X \setminus Y)$. The surface $V = X \setminus Y$ is a homology plane iff this Picard group is zero. Its plane divisor $D(V)$ is $c(Y) = L$ where $c = \pi'\pi$.*

Remarks. The existence of choices in steps 2 and 3 above leads to families of homology planes which by construction have the same original line configuration as plane divisor. These families are parametrized by the values of the edge and vertex functions as well as the selections of exceptional fibers in step 2. It is not the place here to illustrate with examples all the possibilities which can occur in all the above steps in producing homology planes with a given linear plane divisor. For illustrative purposes we demonstrate the procedure in the case where the line configuration is 4 lines in general position in $\mathbf{P}^2$. In that case we select in step 2 all the exceptional divisors to be members of the new configuration and in steps 3 and 4 we need no vertex function—only an edge function. Step 6 requires a computation of $\mathrm{Pic}(X \setminus Y)$ in terms of edge and vertex functions. This is done in 1.7, 2.7(v) and 2.9.

Finally, here is our last main result.

Theorem F. *For each line configuration L in Theorem D, there are infinitely many homology planes which have L as plane divisor.*

The paper is organized as follows: In section 1 we treat the geometric aspects of homology planes. In section 2 we illustrate the steps in proving Theorem E in the case of constructing the homology planes whose plane divisor is $L = L(2)$, i.e., consists of 4 lines in general position in $\mathbf{P}^2$. There are 2 infinite families of affine surfaces having $L(2)$ as plane divisor. One is an infinite family of homology planes. The other is an infinite family of $\mathbf{Q}$-homology planes, i.e., non-singular affine surfaces which are acyclic over the rationals $\mathbf{Q}$. These two infinite families are described in Theorem 2.11 which is the special case of Theorem E dealing with $L(2)$. In section 3 we illustrate the use of Cremona transformations in showing that homology planes have tidy plane divisors. As mentioned, these techniques have been used to convert complicated plane divisors into linear plane divisors and hopefully will play a role in answering the question of whether every homology plane has a linear plane divisor.

Acknowledgement. This seems to be a good point to acknowledge the week of stimulating conversations the second author of this paper had at the University of Osaka with Miyanishi a few months prior to the Rutgers Conference. With him we had many interesting discussions on the topic of homology planes from his algebraic geometric viewpoint and our topological viewpoint. Exchanges like this were the intention of the Rutgers Conference as well.

Section 1

We want to use the fact that homology planes have plane divisors to construct homology planes from a curve $D \subset \mathbf{P}^2$, i.e., we want to realize D as the plane divisor of one or several homology planes. To set up the construction, here is a description of the relation between a homology plane, its plane divisor and the blow-up operations:

Lemma 1.1. *If $D \subset \mathbf{P}^2$ is the plane divisor of a homology plane V, then there is a projective surface X, a curve $Y \subset X$ and a contraction $c : (X, Y) \to (\mathbf{P}^2, D)$ such that the following holds:*

(i) $X \setminus Y = V$.

(ii) *We have*

$$Y = \overline{c^{-1}(D \setminus \Sigma)} \cup \bigcup_{u \in U''} E_u = \overline{c^{-1}(D) \setminus \bigcup_{u \in U'} E_u}$$

where Σ is the exceptional set of c, $\{E_u \mid u \in U\}$ is the set of exceptional divisors of c and U' and U'' are complementary subsets of U.

(iii) *$\beta_2(D) + \#U'' = \beta_2(X)$ where $\#U''$ is the cardinality of U'' and β_2 is the second Betti number.*

In order to use this lemma to produce homology planes from a curve $D \subset \mathbf{P}^2$, we need to know when $X \setminus Y$ is a homology plane in the case where $c : X \to \mathbf{P}^2$ is a contraction and Y is defined by condition (ii) in the preceding lemma. This information is provided in 1.8 below but first we need some preliminaries about the Picard group.

The Picard group of a variety X is denoted by $\operatorname{Pic}(X)$. This is the group of algebraic line bundles over X. It is the same as the divisor class group $\operatorname{cl}(X)$ when X is smooth (which is the only case we use). This means that a divisor determines an element of the Picard group. In fact, we deal only with the case where X is a non-singular surface. Let X be such a surface and let Y be a curve on X. The subgroup of $\operatorname{Pic}(X)$ generated by the irreducible components of Y is denoted by $L(Y)$. The following is a special case of a lemma which may be found in [**H**, Chap. 2, 6.5]. It plays an important role in our discussion.

Lemma 1.2. *Let X be a smooth surface and Y be a curve on X. Then* $\operatorname{Pic}(X \setminus Y) = \operatorname{Pic}(X)/L(Y)$.

A map (i.e., algebraic morphism) $c : X \to W$ induces a homomorphism $c^* : \operatorname{Pic}(W) \to \operatorname{Pic}(X)$ which is easy to describe from the definition of Pic in terms of line bundles. The description is more subtle when Pic is viewed as cl. Since we use this viewpoint, we at least need some properties of the induced map in terms of divisor classes. One qualitative property is this: Let Y' be a curve in W. Then $c^*(Y')$ is a linear combination of the irreducible components of $Y = c^{-1}(Y')$ in $\operatorname{cl}(X) = \operatorname{Pic}(X)$. Here we've abused language and used Y', etc. to denote its own representative in $\operatorname{cl}(W)$. The fundamental case is the following:

Lemma 1.3. *Let X be obtained from W by blowing up $p \in W$ and let $c : X \to W$ be the blow-up map. Then for any curve $Z \subset W$ with proper transform $\tilde{Z}$ and $E = c^{-1}(p)$ we have*

(i) $c^{-1}(Z) = \tilde{Z} \cup E$ *and* $c^*(Z) = \tilde{Z} + E$ *if p is a smooth point of Z.*

(ii) $c^{-1}(Z) = \tilde{Z}$ *if p is not in Z. In this case* $c^*(Z) = \tilde{Z}$.

(iii) $\mathrm{Pic}(X) = c^* \mathrm{Pic}(W) \oplus \langle E \rangle$ *and c^* is injective.*

The free abelian group generated by A_i, $i \in I$ is denoted $\langle A_i, i \in I \rangle$. The abelian group generated by these elements and subject to the relations $R_j, j \in J$ is denoted $\langle A_i, i \in I \mid R_j, j \in J \rangle$.

Let $c : X \to W$ be a contraction, $D \subset W$ a curve and

$$Y = \overline{c^{-1}(D) \setminus \bigcup_{u \in U'} E_u}$$

where $\{E_u \mid u \in U\}$ is the set of exceptional divisors of c and U' is any subset of U. Since $\mathrm{Div}(X)$, the divisor group of X, is a free abelian group generated by the irreducible curves in X, there is a natural projection

$$\pi' : \mathrm{Div}(X) \to \langle E_u, u \in U' \rangle. \tag{1.4}$$

Now specialize to the case where W is $\mathbf{P}^2$. Since $\mathrm{Pic}(\mathbf{P}^2)$ is $\langle L \rangle$ where L is any line in $\mathbf{P}^2$, Lemma 1.3 gives:

Proposition 1.5. $\mathrm{Pic}(X) = \langle \tilde{L}, E_u, u \in U \rangle$ *when* $W = \mathbf{P}^2$.

Since Pic is a quotient of Div and $\mathrm{Pic}(X)$ is free abelian when $W = \mathbf{P}^2$, we get the following statement:

Corollary 1.6. *When $W = \mathbf{P}^2$, π' induces a map*

$$\pi : \mathrm{Pic}(X) \to \langle E_u, u \in U' \rangle.$$

Lemma 1.7. *Again suppose that $W = \mathbf{P}^2$ and $D = \cup_{v \in T} D_v$ where each D_v is a line. Let L be any one of these lines. Then for X and Y as in 1.1*

$$\mathrm{Pic}(X \setminus Y) = \langle E_u, u \in U' \mid \pi c^*(D_v - L), v \in T \rangle.$$

PROOF: Observe that

$$\mathrm{Pic}(\mathbf{P}^2) = \langle D_v, v \in T \mid D_v - D_w, v, w \in T \rangle,$$

and so

$$\mathrm{Pic}(X) = \langle \tilde{D}_v, v \in T, E_u, u \in U \mid c^*(D_v - D_w), v, w \in T \rangle$$

by 1.3. Hence,

$$\mathrm{Pic}(X \setminus Y) = \mathrm{Pic}(X)/L(Y) = \langle E_u, u \in U' \mid \pi c^*(D_v - L), v \in T \rangle$$

because $L(Y)$ is generated by $\{\tilde{D}_v, v \in T, E_u, u \notin U'\}$.

Remarks. We have used c^* to denote both the map on Div and on Pic. To use 1.7, we must describe the map πc^*, i.e., we must determine the coefficients of the $E_u, u \in U'$ which occur in $\pi c^*(D_v)$. We do this in 2.7(v) and 2.6(vi) below for one of the situations of application. Under the assumption $\mathrm{Pic}(W \setminus D) = 0$, one may use 1.3 to show $\mathrm{Pic}(X \setminus Y)$ is generated by $\langle E_u, u \in U' \rangle$. If in addition $W = \mathbf{P}^2$, we have a precise computation of $\mathrm{Pic}(X \setminus Y)$.

The importance of the above discussion about Pic is seen in the following lemma which is used to produce homology planes:

Lemma 1.8 (cf. [**R**]). *Suppose that X is a non-singular surface and $Y \subset X$ a connected curve with smooth irreducible components which pairwise intersect transversely in at most one point. Then $X \setminus Y$ is a homology plane if and only if the following conditions hold:*

(i) $\pi_1(Y) = 0$.
(ii) $\mathrm{Pic}(X \setminus Y) = 0$.
iii) *X is rational.*

PROOF: Suppose the assumptions hold. We show that $X \setminus Y$ is a homology plane. The assumptions on Y imply that its irreducible components are topological $\mathbf{P}^1$'s and that its homology is zero except in dimensions 0 and 2. Since X is rational, it is simply connected and its homology vanishes in dimensions different from 0,2 and 4. Also, the natural map of $\mathrm{cl}(X)$ to $H_2(X)$ is an isomorphism. Since the irreducible components of Y are $\mathbf{P}^1$'s, this map induces an isomorphism of $L(Y)$ with $H_2(Y)$. The quotient of $\mathrm{cl}(X)$ by $L(Y)$ is $\mathrm{cl}(X \setminus Y) = \mathrm{Pic}(X \setminus Y)$ which is 0 by assumption. This implies from the mentioned isomorphisms that $\beta_2(Y) = \beta_2(X)$. Moreover, the map of $\mathrm{cl}(X)$ to $H_2(X)$ induces an isomorphism of $\mathrm{cl}(X \setminus Y)$ with $H_2(X, Y) = H^2(X \setminus Y)$ (because $H_1(Y) = 0$) which therefore vanishes. By Poincaré duality and excision, $H_i(X, Y) = H^{4-i}(X \setminus Y)$. The above information implies that $H^i(X \setminus Y) = 0$ for $i \neq 0$ and this implies that $X \setminus Y$ is affine by [**F**, 2.5] and therefore a homology plane. Running this argument backwards proves the converse.

Remarks 1.9. Conditions (ii) and (iii) can be replaced by the assumptions that $\beta_2(Y) = \beta_2(X)$ and $H^2(X \setminus Y) = 0$. This uses the rationality theorem of [**GS**].

Some of the assumptions of the lemma can be weakened. E.g., the components of Y can be just topological $\mathbf{P}^1$'s, like a cubic with a cusp.

Section 2

A *graph* T here will mean an ordinary graph with weight function Ω which is an integral valued function on vertices of T. If v and w are two vertices of T, we write vw if v and w are linked, i.e., form an edge of T. A graph is called *minimal* if each vertex whose weight is -1 is a *branch point*, i.e., connected to at least 3 other vertices. A *tree* is a graph without loops. To emphasize, a tree always comes with a weight function. A *special graph* τ is a graph with (weight function Ω) and a set function m which assigns to each vertex v of τ a set $m(v)$ of $n = n(v)$ positive rational numbers $\{r_1, \dots, r_n\}$. We note that $m(v)$ may be empty. A *special tree* is a special graph without loops.

A special graph τ is converted to a graph $G\tau = T'$ as follows: At each vertex v of τ add $n(v)$ new branches. A linear tree $B(r_i)$ is attached to the rational number r_i in $m(v)$. One for each rational number in $m(v)$. The association of $B(r)$ to a rational number r goes as follows: Let

$$r = [b_s, b_{s-1}, \dots b_1] = b_s - \cfrac{1}{b_{s-1} - \cfrac{1}{b_{s-2} - \ddots \cfrac{1}{b_1}}}$$

be the continued fraction for r. Set $r' = b_s$. The linear tree $B(r)$ is a tree with $s-1$ vertices v_j for $j = s-1, \dots, 1$, edges $v_j v_{j+1}$ and weights $\Omega(v_j) = -b_j$. Call v_{s-1} the final vertex of $B(r)$. The $B(r_i)$ are attached to v by an edge connecting v to the final vertex of each. Now the vertex v in T' gets a new weight which is

$$\Omega(v) + n(v) - \sum_{i=1}^{n(v)} r'_i. \tag{2.1}$$

Geometrically graphs arise from curves on surfaces. We suppose D is a curve on a surface W and its irreducible components are smooth and intersect transversely pairwise in at most one point, then the weighted dual graph or briefly dual graph $T(W, D)$ is defined. Its vertices are the irreducible components of D and two vertices form an edge if the corresponding components intersect. The weight at a vertex is the self intersection number of the corresponding component. If $T = T(W, D)$ is minimal, it is called the *compactification tree* of $W \setminus D$. If there is a special graph $\tau(W, D) = \tau$ such that $G\tau = T$, we say that $\tau(W, D)$ is the *special compactification tree* for $W \setminus D$. Special compactification trees are more efficient to deal with than compactification trees and carry the same information.

We need the definition of *infinitely near points* on a surface W. This is defined relative to a sequence of repeated blow-ups of W. If a point $p \in W$ is blown up with resulting contraction $c : X \to W$, the points of the exceptional divisor $E = c^{-1}(p)$ are said to be infinitely near to p; so too are the points on the exceptional divisor of a blow-up of X of a point $p' \in E$ and so on. All these points in the various successive blow-ups which lie over p are said to be infinitely near to p.

2.2. Let $D \subset W$ be a curve in a non singular projective surface. We suppose, as usual, that the irreducible components of D are smooth and intersect transversely, that no three irreducible components intersect in one point and two irreducible components intersect in at most one point.

The weighted dual graph $T = T(W, D)$ is a convenient means to keep track of repeated blow-ups of W of points in D and their infinitely near points. In fact, there is the notion of blowing up vertices and edges of a graph which correspond to blowing up a point of W which lies on D. Whether a vertex or edge is blown up depends on the placement of the point p in D which is blown up according to the following

Conditions 2.3.

(i) *p is in a unique irreducible component D_v of D.*
(ii) $p = D_v \cap D_w$.

In the first case $T(X, D') = T'$ is obtained from $T(W, D) = T$ by blowing up the vertex v and in the second case by blowing up the edge vw. Here X is the blow-up of W at p and D' is the inverse image of D under the blow-up map $c : X \to W$.

We now briefly define these combinatorial notions of blow up of vertices and edges. (See also [**GS2**].) The graph T' obtained from T by blowing up the vertex v is obtained from T by adding one new vertex v' which is linked to v and to no other vertices of T'. Its weight is -1 while the new weight of v is the old weight minus 1. Other vertices have unchanged weights. The graph T' obtained from T by blowing up the edge vw is obtained from T by adding one new vertex v' which is linked to v and to w and no other vertices of T'. The new weights of v and w are the old weights minus 1. The weight of v' is -1 and the other weights are unchanged.

2.4. If points of D and their infinitely near points are successively blown up in W, the resulting surface X and curve D' depend on the precise choice of points blown up. Correspondingly, $T(X, D') = T'$ depends on the choice of vertices and edges of $T(W, D) = T$ blown up. The data which records this is a vertex function d and an edge funtion M. The function d tells how many times each vertex is blown up while the function M tells the sequence of edges which are blown up. E.g., the function d which has value 1 on $v \in T$ and 0 on other vertices would instruct us to blow up the vertex v once and no other vertices. The function M assigns to each edge of vw of T integers $M(v, vw)$ and $M(w, vw)$ which satisfy:

(i) $M(v, vw)$ and $M(w, vw)$ are non negative.

(ii) Either both are 0 or both are positive and relatively prime. In case both are positive, this is abbreviated by $M(vw) \neq 0$.

As an example, the function M which is 0 on all edges except vw and has $M(v, vw) = 1 = M(w, vw)$ instructs us to blow up the edge vw once and no other edge. For more complicated edge functions the sequence of edges it selects for blow-up is more complicated and the precise rule for this is here immaterial. What we need to know for this discussion is: Each edge function M determines a surface X and a contraction $c : X \to W$ whose exceptional set lies on D. Moreover, the graph MT produced by the edge blow-ups determined by M is $T(X, D')$ where $D' = c^{-1}(D)$.

We now describe the graph MT and the related special graph $\tau = M'T$. First these definitions: Let vw be an edge of T with $M(vw) \neq 0$. Set

$$(M(v, vw) + M(w, vw))/M(w, vw) = [b_{-r}, b_{-r+1}, \dots, b_{-1}]$$

$$(M(v, vw) + M(w, vw))/M(v, vw) = [b_s, b_{s-1}, \dots, b_1]$$

where r and s are positive integers. Then $L(vw)$ is the linear tree with vertices v_i for $i = -r, -r+1, \dots, s$ and edges $v_i v_{i+1}$. The vertices v_{-r}, v_s and v_0 are specially denoted as v, w and $u(vw)$. In particular, v and w are the ends of $L(vw)$. Of course, r and s are functions of M and the edge vw. This is not indicated in the notation.

2.5. MT is the graph obtained from T by replacing each edge vw with $M(vw) \neq 0$ by $L(vw)$. The new weight function Ω' on MT is defined by

(i) $\Omega' = \Omega$ on the complement of the union of the edges where M is not 0.

(ii) On $L(vw)$:

$$\Omega'(v_i) = -b_i, i \neq -r, s \text{ or } 0,$$

$$\Omega'(v) = \Omega(v) + 1 - b_{-r},$$

$$\Omega'(w) = \Omega(w) + 1 - b_s,$$
$$\Omega'(u(vw)) = -1.$$

(iii) $U(M) \subset T$ consists of the interior vertices of all $L(vw)$ with $M(vw) \neq 0$ and $U'(M) \subset U(M)$ consists of the vertices $u(vw) \in L(vw)$ for each vw with $M(vw) \neq 0$.

2.6. $M'T = \tau$ is the special tree whose vertices are the vertices of T. The edges of τ are the edges of T where each edge vw with $M(vw) \neq 0$ is deleted. The weight function Ω' on τ is the same as the weight function Ω on T. The set function m' on τ is defined as follows:
Let v be a vertex of τ.

(i) If there is no edge vw with $M(vw) \neq 0$, $m'(v) = \emptyset$.
(ii) In the contrary case

$$m'(v) = \{(M(v, vw) + M(w, vw))/M(w, vw) \mid M(vw) \neq 0\}.$$

Going hand in hand with the operation of edge blow-up is the operation of *edge cut.* If T is a tree and M is an edge function on T, then CMT is the tree obtained from MT by deleting each vertex in $U'(M)$. (See 2.5(iii).) This tree is easily seen to be $GM'T$. Sometimes we loosely refer to this operation on T as cutting edges of T even though the effect is first to replace the edges of T to be cut by linear trees and then to disconnect these linear trees at the vertices $u(vw)$ for which the edge vw is cut.

As mentioned, these combinatorial notions of blow-up and cutting are really book keeping devices for geometric operations. Blowing up an edge vw in $T(W, D)$ corresponds to blowing up the point $p = D_v \cap D_w$ and its infinitely near points. This produces a surface X and a contraction $c : X \to W$ with $D' = c^{-1}(D)$ for which $T' = T(X, D')$ is the tree obtained by a single edge blow-up. Cutting this edge corresponds to removing in addition the irreducible component of $c^{-1}(p)$ of self intersection number -1. The purpose of edge cutting is to kill cycles in the fundemental group of D. E.g., a cycle in D produced by a chain of irreducible components of D will be killed by cutting any edge vw in T for which the irreducible components D_v and D_w are members of this chain. The ultimate goal is to achieve $\pi_1(Y) = 0$ for Y defined by 2.7(iv). This is the first of the conditions in 1.8 to make $X \setminus Y$ a homology plane with $c(Y) = D$.

Here is further information on the correspondence between the geometry and combinatorics of these operations. We suppose that $D \subset W$ have the properties 2.2 assumed for the constructions of edge blow-up and edge cutting.

2.7. Let M be an edge function on $T = T(W, D)$. Then there is a unique sequence of blow-ups of type 2.3(ii) on W which produces a contraction $c : X \to W$ such that

(i) The exceptional set Σ of c is $\{D_v \cap D_w \mid M(vw) \neq 0\}$.
(ii) $T(X, D') = MT$ for $D' = c^{-1}(D)$.
(iii) The irreducible components of $E = c^{-1}(\Sigma)$, i.e., the exceptional divisors of c are indexed by $U = U(M)$ and $E = \bigcup_{u\in U(M)} E_u$.
(iv) We have

$$T(X,Y) = CMT(W,D) = GM'T(W,D)$$

for $Y = \overline{c^{-1}(D) \setminus \bigcup_{u\in U'(M)} E_u}$. Or, $Y = \overline{c^{-1}(D\setminus\Sigma)} \bigcup_{u\in U''} E_u$ where $U'' \subset U(M)$ is the complement of $U'(M)$.
(v) The map $\pi' : \mathrm{Div}(X) \to \langle E_u, u = u(vw) \in U'(M)\rangle$ of 1.4 is related to M by

$$\pi' c^*(D_v) = \sum_{w\in T: M(vw)\neq 0} M(v, vw) E_{u(vw)}.$$

(vi) In the case $W = \mathbf{P}^2$, π' induces a map

$$\pi : \mathrm{Pic}(X) \to \langle E_u, u = u(vw) \in U'(M)\rangle$$

by 1.6, and if further each D_v is a line and D_w is denoted by L for some fixed $w \in T$, then by 1.7,

$$\mathrm{Pic}(X \setminus Y) = \langle E_u, u = u(vw) \in U'(M) \mid \pi c^*(D_v - L), v \in T\rangle$$

where $U'(M) = \{u(vw) \mid M(vw) \neq 0\}$.

2.8. To illustrate 2.7 in connection with Theorem F, we show how to produce all homology planes having a plane divisor D consisting of 4 lines in general position in $\mathbf{P}^2$. Let $W = \mathbf{P}^2$ and $D = \bigcup_{i=1}^4 D_i$ where D_i is a line in $\mathbf{P}^2$ for $i = 1, \ldots, 4$. Then $T = T(W, D)$ consists of 4 vertices which we view as the 4 vertices 1,2,3 and 4 of a tetrahedron and T as the graph consisting of the edges of this tetrahedron.

The first step is to cut the edges of T to produce a tree. Up to isomorphism there are 2 distinct ways to cut the edges of T to produce a tree:

(i) Cut the edges 12, 23 and 13 to produce a tree T_1.
(ii) Cut the edges 14, 12 and 23 to produce a tree T_2.

The second step is to choose edge functions M_1 and M_2 which are non zero on the edges in (i) resp. (ii) and zero elsewhere; so the tree T_i is $M_i^!T$ for $i = 1, 2$. E.g., in case (ii), the edge function $M = M_2$ has this description:

(iii) $M(1,12) = u_0$, $M(2,12) = u_1$.
(iv) $M(1,14) = v_0$, $M(4,14) = v_1$.
(v) $M(2,23) = w_0$, $M(3,23) = w_1$ where $u_0, u_1, v_0, v_1, w_0, w_1$ are integers which satisfy 2.4.

For the remainder of this section, we treat case (ii) in detail and so set $M_2 = M$. According to the discussion in 2.7 there is a contraction $c : X \to W$ defined by M, and for

$$Y = \overline{c^{-1}(D) \setminus \bigcup_{M(vw) \neq 0} E_{u(vw)}}$$

we have

$$T(X, Y) = GM'T = CMT.$$

2.9. $M'T$ is the special tree which is linear with four vertices 1, 2, 3 and 4 with edges 13, 34, and 42. Its weight function $\Omega = 1$ and its set function is

$$m(1) = \{(u_0 + u_1)/u_1, (v_0 + v_1)/v_1\}$$

$$m(2) = \{(u_0 + u_1)/u_0, (w_0 + w_1)/w_0\}$$

$$m(3) = \{(w_0 + w_1)/w_0\}$$

$$m(4) = \{(v_0 + v_1)/v_0\}.$$

The map

$$\pi c^* : \mathrm{Pic}(\mathbf{P}^2) \to \langle E_{u(vw)}, M(vw) \neq 0 \rangle$$

in case (ii) has this description by 2.7(v):

$$\pi c^*(D_1) = u_0 E_{u(12)} + v_0 E_{u(14)}$$

$$\pi c^*(D_2) = u_1 E_{u(12)} + w_0 E_{u(23)}$$

$$\pi c^*(D_3) = w_1 E_{u(23)}$$

$$\pi c^*(D_4) = v_1 E_{u(14)}.$$

By Lemma 1.7

$$\mathrm{Pic}(X_2 \setminus Y_2) = \langle E_{u(vw)}, vw = 12, 14, 23 \mid \pi c^*(D_i - D_1)i = 2, 3, 4\rangle$$

This group is zero iff the square of the determinant of the 3 equations above is 1, i.e.,

(2.10) $$(-u_1w_1(v_0 - v_1) - v_1u_0(w_0 - w_1))^2 = 1.$$

Thus $X \setminus Y = V$ is a homology plane iff 2.10 holds and in that case $D(V) = D$.

We summarize this discussion in the following theorem:

Theorem 2.11. *Let D be 4 lines in $W = \mathbf{P}^2$ in general position and $T = T(W, D)$. Let M_1 and M_2 be two edge functions on T which are non zero on the edges of T described by 2.8(i) resp. 2.8(ii). Denote either by M. Then M determines a contraction $c : X \to W$. Set*

$$Y = \overline{c^{-1}(D) \setminus \bigcup_{u \in U'(M)} E_u}.$$

Then $X \setminus Y = V$ is a homology plane iff 2.10 is satisfied in case $M = M_2$, and in that case its special compactification tree $\tau(V) = M'T$ is given in 2.9.

Remarks. In case $M = M_1$ an equation analogous to 2.10 must hold in order that V be a homology plane. This equation never holds as the right hand side is never 1 or -1; however when the right hand side is non zero, V is a rational homology plane whose compactification tree is described in 2.7(iv). In either case the plane divisor of all these homology planes ($\mathbf{Q}$-homology planes) is D. Conversely, every homology plane ($\mathbf{Q}$-homology plane) having D as its plane divisor occurs from some edge function M_1 (M_2).

Section 3

The basic notion used in this paper to study homology planes is *plane divisor*. As we have seen, each homology plane has a plane divisor. For the problem of classification of homology planes, it is important to find a "simple plane divisor" for each homology plane. In 3.6 we show that each homology plane has a plane divisors with tidy singularities.

A plane divisor of a homology plane is however not unique. Two different plane divisors of a homology plane are related by a Cremona transformation, i.e., a birational automorphism of the plane. On the other hand an isomorphism between homology planes defines a Cremona transformation. In this section we also briefly explore these relationships in the context of producing an isomorphism between homology planes whose plane divisors are related by a Cremona transformation.

If $c_i : X \to \mathbf{P}^2$ for $i = 1, 2$ are two contractions, then $c_1 c_2^{-1} = C$ is a birational automorphism of $\mathbf{P}^2$, since contractions are birational isomorphisms. Then, as birational maps $c_1 = Cc_2$. Suppose that $Y \subset X$ is a curve and $X \setminus Y$ is a homology plane. There is a simple description of the relationship between the two plane divisors $D_1 = c_1(Y)$ and $D_2 = c_2(Y)$ and the Cremona transformation C which we now explain. Let S be the exceptional set of c_2 and let D be a curve in W. Then $C(D)$ is defined to be $\overline{C(D \setminus S)}$. Suppose that S lies in D_2. then $Y = c_2^{-1}(D_2 \setminus S) \cup R_2$ where R_2 is a union of exceptional divisors of c_2. Applying c_1 to this equality, we find that

$$D_1 = C(D_2) \cup c_1(R_2) \tag{3.0}$$

In certain situations it is possible to simplify a plane divisor of an affine surface by successive application of Cremona transformations to the plane divisor and by adding or subtracting lines to the successive images as suggested by 3.0. This process can be used to simplify the singularities of the plane divisor. The same technique may sometimes be used to show that two homology planes with plane divisors D_1 and D_2 are isomorphic by producing an isomorphism between them which realizes a Cremona transformation which transforms D_2 to D_1 via 3.0. We first illustrate the simplification of singularities and then touch on the realization of isomorphisms between homology planes whose plane divisors are related by a Cremona transformation at the end.

The group of Cremona transformations is generated by *quadratic transformations*: Let S be a set of 3 points in $\mathbf{P}^2$ which are not collinear and let J be the three lines joining them. The birational transformation of $\mathbf{P}^2$ defined by blowing up these 3 points and blowing down the proper transforms of the lines joining them is called the quadratic transformation $C = C_S$. The points of S are called the *centers* of C and the lines in J called the *lines of* C. Let W be the surface obtained by blowing up the three points. Then W comes equiped with a map to $\mathbf{P}^2$, called c_1, and $\mathbf{P}^2$ is obtained from W by blowing down the 3 lines in $J'' = c_1^{-1}(S)$.

There is another map from W to $\mathbf{P}^2$ obtained by blowing down the proper transforms (under c_1) of the three lines in J. This map is denoted by c_2. The set $J' = c_2(J'')$ consists of 3 lines in $\mathbf{P}^2$ which intersect in a set S' of three non collinear points; so a new quadratic transformation $C_{S'}$ is defined which is in fact the inverse C^{-1} of $C = C_S$. Its center set is S' and line set is J'.

Here are some properties of the quadratic transformation C_S:

3.1. $C = C_S = c_2 c_1^{-1}$ as a birational transformation. Note that C is defined as an honest map on the complement of its center set S and the image of C is the complement of J' the set of lines of C^{-1}. Though C is not defined on S, we set $C(x) = c_2 c_1^{-1}(x)$ for $x \in S$. This is a line in J' and all such lines are obtained this way by definition. The image of a curve Y in $\mathbf{P}^2$ under C is by definition the image under c_2 of the proper transform under c_1 of Y. An equivalent description is that $C(Y)$ is the closure $\overline{C(Y \setminus S)}$ of $C(Y \setminus S)$.

Lemma 3.2. *Let K be a curve in $\mathbf{P}^2$ and let C be a quadratic transformation. If the lines of C intersect K transversely, then the lines of C^{-1} intersect $C(K)$ transversely.*

Remark. Lemma 3.2 is a consequence of the following fact: Let A and B be curves in a smooth surface X. Then A and B intersect transversely at a point p if and only if in the blow-up of X at p, $A' \cap B' \cap E_p = \emptyset$. Here A', B' denote the proper transforms and E_p is the exceptional fiber of the blow-up. To verify Lemma 3.2, apply this fact with A a line of C^{-1} and $B = C(K)$. Remember that C is defined by blowing up its centers and blowing down the proper transforms of its lines.

Lemma 3.4. *Let X' be a non-singular surface and Y' a curve on X'. Let $c' : X' \to \mathbf{P}^2$ be a contraction and let $D = c'(Y)$. If $C = C_S$ is any quadratic transformation of $\mathbf{P}^2$ with $S \subset D$, there is a non-singular surface X, a contraction $c : X \to \mathbf{P}^2$ and a contraction $f : X \to X'$ such that:*

(i) *$f : X \setminus Y \to X' \setminus Y'$ is an isomorphism where $Y = f^{-1}(Y')$.*
(ii) *There is a commutative diagram*

$$\begin{array}{ccc} X & \xrightarrow{\;f\;} & X' \\ {\scriptstyle c}\downarrow & & \downarrow{\scriptstyle c'} \\ \mathbf{P}^2 & \xleftarrow{\;C\;} & \mathbf{P}^2 \end{array}$$

(iii) *The plane divisor* $c(Y) = D'$ *is related to the plane divisor* D *by* $D' = L' \cup C(D)$ *where* L' *is a curve whose irreducible components are some of the lines of* C^{-1}.

PROOF: We construct the following commutative diagram:

$$\begin{array}{ccc} X & \xrightarrow{\ f\ } & X' \\ {\scriptstyle b}\downarrow & & \downarrow{\scriptstyle b'} \\ W & \xrightarrow[f']{} & Z \\ {\scriptstyle d_1}\downarrow & & \downarrow{\scriptstyle e} \\ \mathbf{P}^2 & \xleftarrow{\ C\ } & \mathbf{P}^2 \end{array}$$

In this diagram set $d_1 b = c$ and $eb' = c'$: Let W be obtained from $\mathbf{P}^2$ by blowing up the set S of centers of C and let $d_2 : W \to \mathbf{P}^2$ be the corresponding contraction. Let Z be obtained from $\mathbf{P}^2$ by blowing up the points of S which are also blown up by c' and let $e : Z \to \mathbf{P}^2$ be the corresponding contraction. Since W is a blow-up of Z there is a contraction $f' : W \to Z$ with $ef' = d_2$. The map c' factors through Z. Denote this factorization by $c' = eb'$ where $b' : X' \to Z$. Note that b' is a contraction, i.e., a sequence of blow ups. Let X be obtained from W by the corresponding sequence of blow-ups of W. Then there is a map $f : X \to X'$ and contraction $b : X \to W$ such that $b'f = f'b$. Let $d_1 : W \to \mathbf{P}^2$ be the contraction which blows down the d_2 proper transforms of the lines of C. Then $C = d_1 d_2^{-1}$ as a birational transformation. Define $c : X \to \mathbf{P}^2$ to be the composition $d_1 b$.

The assertions (i)–(iii) are now easily verified. By construction, X is obtained from X' by successively blowing up points of X' on Y'; since in addition, Y is the inverse image of Y', the isomorphism (i) is obvious. Assertion (ii) also is clear from the construction. Assertion (iii) follows from 3.0 by letting W there be $\mathbf{P}^2$ and letting $c_2 = d_2 b$ and $c_1 = c = d_1 b$.

Remarks 3.5. A line of C^{-1}, i.e., $C(x)$ for some x in S, is a component of L' if either the point x is not blown up by the contraction c' or if x is blown up by c', then the b' proper transform of $e^{-1}(x)$ is a component of Y'.

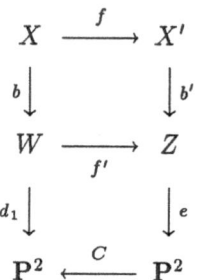

Corollary 3.6. *Let* V *be a nonsingular affine rational surface which has a plane divisor* D. *Then there is a plane curve* K *and a birational transformation* T *of* $\mathbf{P}^2$ *which is a composition of quadratic transformations*

such that $T(D) \cup K$ is a plane divisor of V and this plane divisor has only tidy singularities.

PROOF: By Max Noether [**BK**], there is a sequence of quadratic transformations which converts D to a curve with only tidy singularities. We can arrange that the centers of these quadratic transformations lie on the successive images of D and in addition the lines of each quadratic transformation in this sequence are transverse to the preceding image of D. This involves a general position argument. Let C be the first quadratic transformation in this Noether sequence. By Lemma 3.4, $C(D) \cup L'$ is a plane divisor of V where L' is some subset of the lines of C^{-1}. By Lemma 3.2 the lines of C^{-1} are transverse to $C(D)$ since the lines of C are transverse to D by assumption. This means that the singularities of $C(D) \cup L' = D''$ which are not tidy are already untidy singularities of $C(D)$. Now apply the second quadratic transformation in the Noether sequence for D to D'' and add the neccessary lines to the image to make it a plane divisor of V. Repetition of this process produces the required plane divisior D'.

The problem of classifying homology planes seems formidable. In Theorems D, E and F we have indicated how homology planes arise from configurations of lines in $\mathbf{P}^2$ thus realizing these configurations as plane divisors of homology planes. There then arises the question of whether some of these homology planes with distinct linear plane divisors are sometimes isomorphic and whether some are isomorphic to other homology planes constructed in the literature. This is indeed the case and the group of Cremona transformations is crucial in analyzing this question. Here is one of many results in that direction.

Theorem 3.7 ([tDP2]). *The Ramanujam homology plane has a linear plane divisor.*

This means that the Ramanujam homology plane arises from those constructed from Theorems D and E.

Let D be the configuration in $\mathbf{P}^2$ consisting of the cubic and conic used to define the Ramanujam homology plane. Let L be the line configuration in $\mathbf{P}^2$ consisting of 5 lines having 2 points of intersection of valence 3 and other intersection points of valence 2. The theorem is proved by showing that there is a Cremona transformation C converting D to L (as in 3.0) which is realized by an isomorphism between the Ramanujam homology plane and one of the homology planes constructed in Theorems D and E with the plane divisor L .

References

[**BK**] Brieskorn E., and H. Knörrer, "Plane algebraic curves," Birkhäuser-Verlag, Boston, 1986.

[**tDP**] tom Dieck, T., and T. Petrie, *Contractible affine surfaces of Kodaira dimension 1*, to appear in J. Math. Soc. Japan.

[**tDP2**] tom Dieck,T., and T. Petrie, *Homology planes and algebraic curves, I*, Preprint.

[**F**] Fujita, T., *On the topology of non-complete surfaces of Kodaira dimension 1*, J. Fac. Sci. Univ. Tokyo **29** (1982), 503–566.

[**GM**] Gurjar, R.V. and M. Miyanishi, *Affine surfaces with* $\bar{\kappa} \leq 1$, in "Algebraic Geometry and Commutative Algebra in Honor of Masayoshi Nagata," 1987, pp. 99–124.

[**GS1**] Gurjar, R.V., and A.R. Shastri, *On the rationality of complex homology 2 cells*, Preprint.

[**GS2**] Gurjar, R.V., and A.R. Shastri, *The fundamental group at infinity of affine surfaces*, Comment. Math. Helv. **60** (1984), 459–484.

[**H**] Hartshorne, R., "Algebraic Geometry," Springer-Verlag, New York, 1977.

[**I**] Iitaka, S., "Algebraic Geometry," Springer-Verlag, New York, 1982.

[**K**] Kraft, H., *Algebraic automorphisms of affine space*, In this volume.

[**M**] Miyanishi, M., *Examples of homology planes of general type*, Preprint.

[**P**] Petrie, T., *Algebraic automorphisms of affine surfaces*, Invent. Math., to appear.

[**R**] Ramanujam, C.P., *A topological characterization of the affine plane as an algebraic variety*, Ann. of Math. **94** (1971), 69–88.

Tammo tom Dieck
Mathematisches Institut
Universität Göttingen
Bunsenstr. 3–5
D-3400 Göttingen
WEST-GERMANY

Ted Petrie
Department of Mathematics
Rutgers University
New Brunswick, NJ 08903
USA

FIXED POINT FREE ALGEBRAIC ACTIONS ON VARIETIES DIFFEOMORPHIC TO $\mathbf{R}^n$

KARL HEINZ DOVERMANN, MIKIYA MASUDA
AND TED PETRIE[1]

1. Introduction

Let $\mathbf{F}$ be the field $\mathbf{R}$ of real or the field $\mathbf{C}$ of complex numbers. This paper deals with the

Fixed Point Problem (see [**PR1**]). *Does an algebraic action of a reductive group on affine space $\mathbf{F}^n$ have a fixed point?*

In order to treat this problem there are two obvious properties of the situation which might be crucial:

(1) The homological property that $\mathbf{F}^n$ is acyclic.
(2) The categorical property of the action: It is algebraic and the group is reductive (e.g., the group is finite).

Do these properties individually or jointly imply the existence of a fixed point? Certainly they play a role. For example, any cyclic group G not of prime power order acts smoothly on $\mathbf{R}^n$ without fixed points, where n depends on G (see [**B**, pp. 58–62]). On the other hand Petrie and Randall [**PR1**] show that a cyclic group acting real or complex algebraically on affine space has a fixed point. The proof uses smooth methods, and it also applies to varieties diffeomorphic to affine space. So the Fixed Point Problem is also interesting in this more general setting of real or complex algebraic actions on varieties diffeomorphic to affine space. In this paper we will concentrate on the real algebraic case.

Let $p_1, \ldots, p_k : \mathbf{R}^n \to \mathbf{R}$ be polynomials. The set of common zeros of these polynomials

$$V = \{x \in \mathbf{R}^n \mid p_j(x) = 0 \text{ for } 1 \leq j \leq k\} \subset \mathbf{R}^n$$

[1]Partially supported by an NSF Grant.

is called a *real algebraic variety*. Suppose G is a compact Lie group and G acts orthogonally on $\mathbf{R}^n$. If G acts on the variety V such that the action is the restriction of the action on $\mathbf{R}^n$, then we call V, together with this action, a *real algebraic G-variety*. We also say that G acts real algebraically on the variety V.

Without going into the motivation, here is a conjecture which suggests the answer to the Fixed Point Problem.

Fixed Point Conjecture. *A compact Lie group which acts smoothly on a disk without fixed points acts real algebraically without fixed points on a variety which is diffeomorphic to $\mathbf{R}^n$ for some n.*

The word disk refers to a compact disk. The converse of this conjecture has been shown in [**PR1**] for $n \geq 5$. Among the groups which act smoothly on a disk without fixed points are those which act smoothly on a homotopy sphere with the property that there is just one fixed point. (In this paper homotopy sphere means smooth closed manifold homotopy equivalent to a sphere.) For this class of groups we have a method for proving the conjecture. We illustrate this method in the case of the icosahedral group $I = A_5$. Aside from the ideas set forth here, the main result is Corollary 1.2. It is a consequence of Theorem 1.1 (below) and the lemma following the theorem.

Corollary 1.2. *For any integer $k \geq 24$, there exists an effective fixed point free real algebraic action of the icosahedral group on a variety diffeomorphic to $\mathbf{R}^k$.*

Let $\mathcal{G}$ denote the set of all compact Lie groups which act real algebraically, effectively and without fixed point on a variety diffeomorphic to affine space. In Section 6 we show

(1) If $H \subset G$ is a subgroup of finite index and $H \in \mathcal{G}$ then $G \in \mathcal{G}$.
(2) If G surjects onto K and $K \in \mathcal{G}$ then $G \in \mathcal{G}$.

This, together with Corollary 1.2, implies that there are infinitely many groups which act real algebraically on a variety diffeomorphic to affine space without a fixed point.

Theorem 1.1. *For any integer $k \geq 2$, there is an effective real algebraic action of the icosahedral group with exactly one fixed point on a variety which is diffeomorphic to a homotopy sphere of dimension $12k$.*

Lemma. *Any real algebraic action of a compact Lie group G on a variety which is a (smooth) homotopy sphere of dimension n and which has exactly one fixed point gives rise to a fixed point free real algebraic action of G on a variety diffeomorphic to $\mathbf{R}^n$.*

PROOF: (Compare [**M1**, p. 105].) Let $(p_1, \ldots, p_k)$ be the ideal which defines the variety V on which G acts with exactly one fixed point. Let Ξ denote the representation of G in which V is the zero set. Set $p = \sum p_i^2$. Then $p^{-1}(0) = V$. Find an invariant polynomial q such that $q^{-1}(0)$ is the fixed point. Then

$$V \setminus V^G = \{x \in \Xi \mid p(x) = 0 \text{ and } q(x) \neq 0\}.$$

The assignment which maps x to $(x, 1/q(x))$ defines an equivariant diffeomorphism

$$V \setminus V^G \to A = \{(x, y) \in \Xi \oplus \mathbf{R} \mid p(x) = 0 \text{ and } yq(x) - 1 = 0\}$$

The action of G on A is real algebraic. Finally, observe that A is diffeomorphic to a homotopy sphere with one point removed. This implies that A is diffeomorphic to $\mathbf{R}^n$. ∎

The strategy for proving the Fixed Point Conjecture for groups which act smoothly on a homotopy sphere with exactly one fixed point is based on the Equivariant Nash Conjecture. Following Akbulut and King [**AK2**] we state its relative version. Here G denotes a compact Lie group.

Relative Equivariant Nash Conjecture. *Let M be a closed smooth G-manifold, and let M_i, $i = 1, \ldots, k$, be closed smooth G-submanifolds in general position. Then there is a real algebraic G-variety X and closed algebraic G-subvarieties X_i of X and a G-diffeomorphism $\Phi: M \to X$ with $\Phi(M_i) = X_i$.*

Unfortunately, the Equivariant Nash Conjecture has not been proved. However, we can prove a version which makes an equivariant bordism assumption. Let M_1 and M_2 be two smooth closed manifolds with smooth actions of a compact Lie group G. They are called *G-cobordant* if there exists a smooth G-manifold W such that its boundary is the disjoint union of M_1 and M_2.

Theorem 1.3. *Suppose G is a compact Lie group, and M is a closed smooth G-manifold. If M is G-cobordant to a real algebraic G-variety, then M is G-diffeomorphic to a real algebraic G-variety.*

In Sections 4 and 5 we prove Theorem 1.3. The content of these sections corresponds roughly to Sections 1 and 2 in a paper by Akbulut and King [**AK1**]. There they prove the non-equivariant Nash Conjecture. We generalize their arguments. Some arguments are simplified because our Theorem 1.3 generalizes only part of their Theorem 2.8. The important difference is that they do not have to make the bordism assumption of Theorem 1.3. By a result of Milnor [**M2**] every closed smooth manifold is cobordant to a real algebraic variety.

There are two methods for proving the Fixed Point Conjecture for groups which act on a homotopy sphere with exactly one fixed point. The first one begins with a verification of the absolute ($k = 0$) equivariant Nash Conjecture. Let G be a compact Lie group. One needs to show that the algebraic generators of the equivariant bordism ring of G are realized by real algebraic G-varieties. This will imply that every closed smooth G-manifold is equivariantly cobordant to a real algebraic G-variety. Together with Theorem 1.3 it will imply the absolute equivariant Nash Conjecture. Applied to one fixed point actions on spheres, it implies the Fixed Point Conjecture for those groups which act smoothly on a homotopy sphere with exactly one fixed point.

The second method also uses Theorem 1.3, but it does not depend on a solution of the equivariant Nash Conjecture. This is the method used in this paper.

Outline of the Proof of Theorem 1.1: Let $\Sigma = SO(3)/I$ be the Poincaré homology sphere. The icosahedral group I acts on Σ with exactly one fixed point, and with this action Σ is equivariantly diffeomorphic to a real algebraic I-variety. We discuss Σ and its properties in detail in Section 2. Let Σ_{4k} be the $4k$-fold cartesian product of Σ with itself. The icosahedral group acts again real algebraically on Σ_{4k} with exactly one fixed point. In Section 3 (see Theorem 3.1) we use equivariant surgery to construct an equivariant cobordism between Σ_{4k} and a homotopy sphere such that the action on this homotopy sphere has exactly one fixed point. Then Theorem 1.3 implies that the action on the homotopy sphere is equivariantly diffeomorphic to a real algebraic action on a variety. ∎

We now turn to the history related to our results. The classical Nash Conjecture (suppose that G is the trivial group and that $k = 0$) was supported by results of Seifert [**Se**] and Nash [**N**]. Nash made the conjecture in the paper referenced. Tognoli proved this conjecture [**T**]. Several other interesting proofs and generalization have been published since by King [**K**], Akbulut and King [**AK1**] and [**AK2**], Ivanov [**I**], and Bochnak, Coste, and Roy [**BCR**]. All of these proofs of the Nash Conjecture, at least in the absolute case, have two major steps. First one shows that every closed manifold which is cobordant to a variety is also diffeomorphic to a variety. The second step shows that every cobordism class is represented by a variety. Our Theorem 1.3 generalizes the first of these steps to the equivariant setting.

The question of one fixed point actions on homotopy spheres was raised by Montgomery and Samelson [**MS**] in 1946. Our Theorem 1.1 provides the first actions of this type in the real algebraic category. Such actions were constructed in the smooth category by Stein [**St**] for $\mathrm{SL}(2, \mathbf{Z}_5)$ (the binary icosahedral group) and Petrie [**P1**], [**P2**] for odd order abelian groups with at least three non-cyclic Sylow subgroups, S^3, $\mathrm{SO}(3)$, $\mathrm{SL}(2, \mathbf{F})$ and $\mathrm{PSL}(2, \mathbf{F})$ for $\mathbf{F}$ with characteristic ≥ 5. Morimoto [**Mo1**] constructed one fixed point actions of the icosahedral group on S^6. No such actions exist in dimensions ≤ 4 (see [**F**] and [**Mo2**]) or in dimension 5 [**BKS**]. From the construction of Stein, Petrie, and Morimoto it is not clear that the actions are equivariantly cobordant (and with this by Theorem 1.3 equivariantly diffeomorphic) to real algebraic actions.

Our Corollary 1.2 addresses the classical question of fixed point free actions on $\mathbf{R}^n$ and provides the first actions of this kind in the real algebraic category. But, we only know that the underlying space is diffeomorphic to $\mathbf{R}^n$ and not that it is real algebraically isomorphic to $\mathbf{R}^n$. The first fixed point free actions on $\mathbf{R}^n$ were constructed by Conner and Floyd [**CF**] for cyclic groups of order pq where p and q are relatively prime and greater than one. See also [**B**, pp. 58–62]. The complete answer in the smooth category is based on results by Conner and Montgomery [**CM**], by Hsiang and Hsiang [**HH**], and by Edmonds and Lee [**EL**]. In the following theorem 1 is considered a prime power.

Theorem. *A compact Lie group G (with connected component G_0) has a fixed point free smooth action on some $\mathbf{R}^n$ if and only if G/G_0 is not of prime power order or G_0 is not abelian.*

Petrie and Randall [**PR1**] consider the Fixed Point Problem in the real and complex algebraic setting. They cite several references which show that particular algebraic actions on $\mathbf{F}^n$ must have fixed points. They also show this for groups G which have a normal series $P < H < G$ where P and G/H are of prime power order, and H/P is cyclic. A smooth action of such a group on a disk must have a fixed point. A fixed point free complex algebraic action of a reductive group on $\mathbf{C}^n$ would be a striking counter example to Kambayashi's Linearity Conjecture [**Ka**]. The conjecture says that reductive group actions on $\mathbf{C}^n$ are algebraically conjugate to linear actions. For some results concerning this conjecture see also the articles by Bass and Haboush [**BH**] and Kraft [**Kr**]. Recently G. Schwarz showed that this conjecture is false [**S2**], but the actions which he constructs have fixed points.

Much of this work was done during a stay of Mikiya Masuda and Ted Petrie in Hawaii, and much of the final write up was done during a stay of Karl Heinz Dovermann and Mikiya Masuda at the Sonderforschungsbereich in Göttingen. The authors would like to thank both institutions for their hospitality.

2. Study of SO(3)/I

Let I denote the icosahedral group. Its elements are the rotations of the icosahedron, and as such we consider I as a subgroup of SO(3). Both, SO(3) and I act on the left coset space

$$\Sigma = \mathrm{SO}(3)/I = \{gI \mid g \in \mathrm{SO}(3)\} \tag{2.1}$$

by left translation. These are smooth actions which preserve orientations. The manifold Σ is the well-known Poincaré homology sphere.

Proposition 2.2. *There exists a real algebraic I-variety V which is I-diffeomorphic to Σ with I-action by left translation.*

To prove this statement we recall the quotient map construction discussed by G. Schwarz [**S1**]. Let Ω be an orthogonal representation of a compact Lie group G. The algebra of G-invariant polynomials on Ω is finitely generated. Let $\rho_1, \ldots, \rho_k$ be generators, and let $\rho = (\rho_1, \ldots, \rho_k)$ denote the corresponding map from Ω to $\mathbf{R}^k$. This function ρ is a proper map, and ρ induces a homeomorphism between Ω/G and the closed semi-algebraic subset $\rho(\Omega)$ of $\mathbf{R}^k$. In particular, ρ separates orbits, and every orbit is a real algebraic G-variety.

PROOF OF PROPOSITION 2.2: By the Mostow-Palais Embedding Theorem [**B**, p. 111] we may embed the smooth SO(3)-manifold Σ in some orthogonal representation Ω of SO(3), and its image is exactly one SO(3)-orbit. By the argument from above, it is a real algebraic SO(3)-variety. With the restricted action it is a real algebraic I-variety which has all of the properties demanded in the proposition. ∎

Next we discuss the subgroups of the icosahedral group I. Observe that the subgroup A_4 is isomorphic to the tetrahedral group T with 12 elements. Denote the cyclic group of order m by C_m and the dihedral group of order $2m$ by D_{2m}. We arrange the conjugacy classes of subgroups of I in Diagram 2.3. In Table 2.4 we list the normalizers of the subgroups of A_5, and in Proposition 2.5 we summarize information on the H-fixed point sets of Σ, $H \subset I$. This information is helpful for many of the following arguments.

Diagram 2.3.

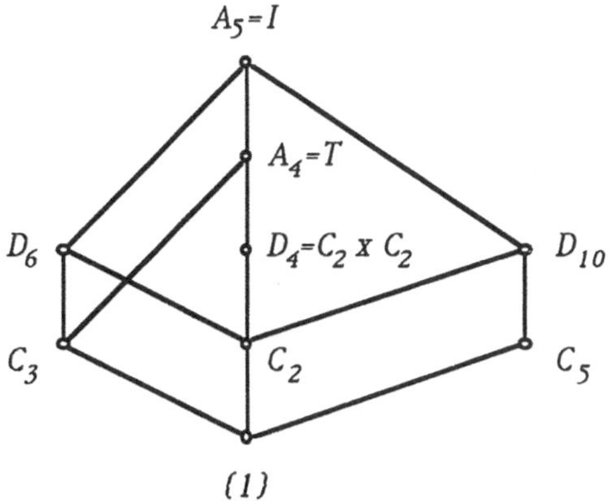

Table 2.4.

(1) $N_I T = T$ and $N_I D_{2m} = D_{2m}$ for $m = 3$, 5.
(2) $N_I D_4 = T$ and $N_I D_4/D_4 = C_3$
(3) $N_I C_m = D_{2m}$ and $N_I C_m/C_m = C_2$

Proposition 2.5.

(1) $\Sigma^I = \{x\}$ *(one point)*
(2) $\Sigma^{C_m} = S^1$ *for* $m = 2$, *3, and 5*
(3) $\Sigma^{D_{2m}} = S^0$ *for* $m = 2$, *3, and 5*
(4) $\Sigma^T = \Sigma^{D_4}$
(5) $\dim(\Sigma^{C_m} \cap \Sigma^{C_{m'}}) = 0$ *for* $m \neq m'$.

Next we study the equivariant tangent bundle of $\Sigma = \mathrm{SO}(3)/I$. Let Ξ be the tangent representation of Σ at the fixed point. This representation is also described by the Lie algebra of SO(3). For a compact Lie group G and a finite subgroup Γ one has $T(G/\Gamma) \cong G \times_\Gamma \mathrm{Lie}(G) \cong G/\Gamma \times \mathrm{Lie}(G)$, and these isomorphisms are G-equivariant with respect to the left action of G on G/Γ. Here $\mathrm{Lie}(G)$ denotes the Lie algebra of G.

Proposition 2.6. *The equivariant tangent bundle $T\Sigma$ of Σ is a product bundle with fibre Ξ, so $T\Sigma \cong \Sigma \times \Xi$.*

PROOF OF 2.5: The arguments are as follows:

(1) $(\mathrm{SO}(3)/I)^I = N_{\mathrm{SO}(3)}I/I = \{1\}$.
(2) The action of C_m on $\mathrm{SO}(3)/I$ is orientation preserving, and thus the C_m fixed set has even codimension. Because there is a fixed point, and C_m does not act trivially, $(\mathrm{SO}(3)/I)^{C_m}$ must be one

dimensional. By Smith theory it is a homology sphere, and thus a circle.

(3) The action of I on Ξ (see 2.6) is the standard action given by the inclusion $I \to \mathrm{SO}(3)$, and the D_{2m} fixed point set is zero dimensional. Thus $(\mathrm{SO}(3)/I)^{D_{2m}}$ is a collection of points. It is also $((\mathrm{SO}(3)/I)^{C_m})^{NC_m/C_m} = (S^1)^{C_2}$. So it must consist of two points.

(4) $(\mathrm{SO}(3)/I)^T = ((\mathrm{SO}(3)/I)^{D_4})^{ND_4/D_4} = (S^0)^{C_3} = S^0$.

(5) This is an immediate consequence of the above. ∎

We will also need a different stable trivialization of $T\Sigma$ than the one given in 2.6. We define it. Consider the double covering $\pi : S^3 \to \mathrm{SO}(3)$. An element $g \in S^3 = S(\mathbf{H})$ is a unit quaternion and an element $\pi(g) \in \mathrm{SO}(3)$ acts on $\mathbf{H} = \mathbf{R} \times \Xi$ by $\pi(g)(\xi) = g\xi g^{-1}$. The representation Ξ consists of the quarternions with vanishing real part. This is the description of Ξ as representation of $\mathrm{SO}(3)$ given in [**B**, p. 56].

Let $c : \Sigma \to S(\mathbf{H})$ be the Thom-Pontrjagin collapse. This map collapses the complement of a small disk around the fixed point x to the point at infinity in $T_x\Sigma^+ = S(\mathbf{R} \times \Xi) = S(\mathbf{H})$. The identification is chosen such that $0 \in T\Sigma$ maps to $1 \in \mathbf{H}$. So $-1 \in \mathbf{H}$ corresponds to the compactification point. The collapse map c is equivariant. Then $c(x) = 1$. If $x \neq y \in \Sigma^H$ for one of the dihedral subgroups H of I, then $c(y) = -1$. Consider the bundle isomorphism

$$\psi : \Sigma \times \mathbf{H} \to \Sigma \times \mathbf{H} \quad \text{defined by} \quad \psi(z, u) = (z, uc(z)).$$

This map is $\mathrm{SO}(3)$-equivariant. Let φ be a bundle isomorphism as obtained in 2.6. Consider the composition

$$\textbf{(2.7)} \qquad B : (\Sigma \times \mathbf{R}) \oplus T\Sigma \xrightarrow{1 \oplus \varphi} \Sigma \times (\mathbf{R} \oplus \Xi) \xrightarrow{\psi} \Sigma \times (\mathbf{R} \oplus \Xi)$$

This bundle isomorphism defines a stable equivariant trivialization of $T\Sigma$. It has one important property which we will use later. Let x and y be distinct points in Σ^H for $H = T$ or $H = D_{2m}$ with $m = 2$, 3, or 5. Consider the restriction of B to the H fixed point set over x and y. By definition (see the values for $c(x)$ and $c(y)$ above)

Lemma 2.8. *The isomorphism* $B^H_{|y} \circ (B^H_{|x})^{-1} : \mathbf{R} \to \mathbf{R}$ *is multiplication with* -1.

3. Proof of Theorem 1.1

In Section 2 we studied the smooth I-manifold Σ. In particular, Σ is diffeomorphic to a real algebraic I-variety (2.2) and the action has exactly one fixed point (2.5(1)). Let Σ_k denote its k-fold cartesian product. Obviously, Σ_k is a real algebraic I-variety and its fixed point set consists of exactly one point. We show

Theorem 3.1. *For every $k \geq 2$, Σ_{4k} is I-cobordant to a homotopy sphere with smooth action of I with exactly one fixed point.*

PROOF OF THEOREM 1.1: Theorem 3.1 provides a homotopy sphere M of dimension $12k$. It has a smooth I-action with exactly one fixed point. Furthermore, it is equivariantly cobordant to a real algebraic I-variety. By Theorem 1.3, M is diffeomorphic to a real algebraic I-variety. Thus M is as required in Theorem 1.1. ∎

More generally than in Theorem 3.1 we will provide a list of conditions for a smooth I-manifold X which guarantee that it is equivariantly cobordant to a homotopy sphere relative to the fixed point set. To state the conditions we need to introduce some topological concepts.

Let G be a compact Lie group. Let Ω be a representation of G. The product bundle with fibre Ω is denoted by $\underline{\Omega}$. The base space will be understood from context.

Suppose the H fixed point sets Ω^H are oriented for every subgroup H of G. Each $g \in W(H) = NH/H$ induces a map of Ω^H to itself, and we assume that these maps preserve the orientation of Ω^H. We call Ω together with the orientations of the H fixed sets and the orientation preserving actions of $W(H)$ a *G-oriented representation.* Similarly, a smooth G-manifold X is called *G-oriented* if each component of X^H is oriented and the induced action of $g \in W(H)$ on X^H preserves this orientation. Let $x \in X$, let $G_x = \{g \in G \mid gx = x\}$ be its isotropy group, and suppose that X is G-oriented. Then the tangent space T_xX has an induced G_x-orientation. Conversely, if we are given a G-vector bundle isomorphism $\varphi : TX \oplus \underline{\Gamma} \to X \times \Omega$, and Ω and Γ are G-oriented representations, then X inherits a G-orientation.

For G-oriented manifolds we have a natural concept of cobordism, called G-oriented cobordism. Two closed G-oriented manifolds M_0 and M_1 are G-orientedly cobordant if there exists a G-oriented manifold W whose boundary is $M_0 \sqcup -M_1$. Here $-M$ is the G-oriented manifold obtained from M by reversing the orientations of all components of M^H for all $H \subset G$. Let H be a subgroup of G. A cobordism is said to be relative to the H fixed point set if $W^H = M_0^H \times I$. In this case we call M_0 and M_1 cobordant relative to the H fixed point set.

Next we discuss the equivariant signature. The principal reference is [**AS**, Section 6]. Our treatment is taken out of [**PR1**, Chapter 2, Section 6]. Let X be a smooth G-oriented manifold on which G acts such that the action preserves the G-orientation. Let H be a subgroup of G. Suppose $\dim X^H$ is defined (i. e., the dimension of a component of X^H depends only on H), and $\dim X^H = 4k$. Consider the middle dimensional cohomology group $H^{2k} := H^{2k}(X^H, \mathbf{R})$ and the intersection pairing $B : H^{2k} \times H^{2k} \to \mathbf{R}$. Here $B(x, y)$ is defined as $(x \cup y)[X^H]$, which is the cup product of x and y evaluated on the fundamental class of X^H. The form B is symmetric. The group $W(H)$ acts on H^{2k}. There is a $W(H)$ invariant positive definite inner product $< , >$ on H^{2k}. Define a linear map $A : H^{2k} \to H^{2k}$ by $B(x, y) =< x, Ay >$. Then $A^t = A$. Write $H^{2k} = H^+ \oplus H^-$, where H^+ (respectively H^-) is the positive (respectively negative) eigenspace of A. These eigenspaces are G-invariant and give characters ρ^+ and ρ^- in $\mathrm{RO}(G) \subset R(G)$. We define

$$\mathrm{sign}(W(H), X^H) = \begin{cases} \rho^+ - \rho^- \in \mathrm{RO}(G) & \text{if } \dim X^H \equiv 0 \pmod 4 \\ 0 & \text{if } \dim X^H \equiv 1 \pmod 2 \end{cases}$$

The signature is also defined if $\dim X^H \equiv 2 \pmod 4$, but we do not need the definition in our context.

Here are the key properties of the equivariant signature which we will use in this paper.

Proposition 3.2. *Let X and Y be smooth G-manifolds which are G-oriented. Then*

(1) $\mathrm{sign}(W(H), X^H) = \mathrm{sign}(W(H), Y^H)$ *for all $H \subset G$ if X and Y are G-equivariantly cobordant as G-oriented manifolds. In particular, if X^H bounds $W(H)$ equivariantly, then* $\mathrm{sign}(W(H), X^H) = 0$.

(2) $\mathrm{sign}(W(H), X^H) = 0$ *if X^H is a homotopy sphere of positive dimension.*

(3) $\mathrm{sign}(W(H), (X \times Y)^H) = \mathrm{sign}(W(H), X^H) \otimes \mathrm{sign}(W(H), X^H)$.

Next we define the notion of a stable G-manifold. Let Ω be a representation of a finite group H. We can write $\Omega = \sum a_\chi \chi$, where χ runs through the irreducible representations of H. The multiplicity $m_\chi(\Omega)$ of the irreducible representation of χ in Ω is a_χ. The module of linear H endomorphisms of χ is a division algebra (by Schur's Lemma) whose dimension over $\mathbf{R}$ is denoted by d_χ. Let 1 denote the trivial representation of H. We say that Ω is H *stable* if for each irreducible representation χ we have

$$m_1(\Omega) \le d_\chi m_\chi(\Omega) \quad \text{if} \quad m_\chi(\Omega) \ne 0.$$

A smooth G-manifold X is called G-*stable* if the tangent representation $T_x X$ is G_x-stable for each $x \in X$.

There are two sets of assumptions which we impose. The first set is used in the process of equivariant surgery in general, and the second one is used specifically in the process of zero-dimensional surgery. They apply to a closed (i.e., compact and without boundary) smooth G-manifold X.

Hypothesis 3.3.

(1) *Suppose for all* $H \subset G$ *all components of* X^H *have the same dimension,* $\dim X^H \equiv 0 \pmod 4$ *and* $\dim X^H \neq 4$.
(2) *If* $H \subset K \subset G$ *and* $\dim X^H \neq \dim X^K$ *then* $2 \dim X^K < \dim X^H$.
(3) X *is stable.*

Hypothesis 3.4.

(1) X^G *consists of exactly one point.*
(2) *There exists a* G*-oriented representation* Ω *and a* G*-vector bundle isomorphism* $b : \underline{\mathbf{R}}^k \oplus TX \to \underline{\mathbf{R}}^k \oplus \underline{\Omega}$.
(3) *If* $H \subset G$ *is an isotropy group of the action on* X, $H \neq G$, *and* $\dim X^H = 0$, *then* $NH/H = 1$.
(4) *The set* A *of points of isotropy type* H *as in (3) decomposes as a disjoint union of three sets* A_+, A_-, *and* A_0, *where* $|A_+| = |A_-|$ *and* $|A_0| = 1$.
(5) $X^H = A \sqcup X^G$ *for* H *as in (3) and* A *as in (4).*
(6) *Let* H *be as in (3),* x *any point in* A_+, *and* y *any point in* A_-. *Then the composition* $b^H_{|y} \circ (b^H_{|x})^{-1} : \mathbf{R}^k \to \mathbf{R}^k$ *has a negative determinant.*

Theorem 3.5. *Let* X *be a closed smooth* I*-manifold which is* I*-oriented. Suppose the hypotheses in 3.3 and 3.4 are satisfied and* $\operatorname{sign}(W(H), X^H) = 0$ *for all* $H \subset I$ *for which* $\dim X^H > 0$. *Then* X *is* I*-equivariantly cobordant relative to the fixed point set to a smooth* I*-manifold* Σ *where* Σ^H *is a homotopy sphere for all* $H \neq I$.

PROOF OF THEOREM 3.1: We verify that the hypotheses in 3.3, 3.4, and the assumptions in Theorem 3.5 apply to the manifold Σ_{4k} considered in Theorem 3.1. Then Theorem 3.1 is implied by Theorem 3.5.

The first two conditions in 3.3 are trivial to check. Condition (3) follows easily from Proposition 2.6 based on the knowledge concerning the representation Ξ of I.

We continue with the discussion of Hypothesis 3.4. Observe that Σ_{4k} has exactly one fixed point because Σ^I consists of exactly one point. Let $\varphi: T\Sigma \to \Sigma \times \Xi$ be a bundle isomorphism as obtained in 2.6, and let B be the isomorphism in 2.7. Observe that $T\Sigma_{4k}$ is the $4k$-fold cartesian product of $T\Sigma$. We set

$$b = B \times \varphi \times \cdots \times \varphi : (\underline{\mathbf{R}} \oplus T\Sigma) \times T\Sigma \times \cdots \times T\Sigma \to (\underline{\mathbf{R}} \oplus \underline{\Xi}) \times \underline{\Xi} \times \cdots \times \underline{\Xi}.$$

Let Ω be the $4k$-fold direct sum of Ξ. Then $b : \underline{\mathbf{R}} \oplus T\Sigma_{4k} \to \underline{\mathbf{R}} \oplus \underline{\Omega}$. This is the bundle isomorphism required in 3.4 (2).

The only proper subgroups H of I for which $\dim \Sigma_{4k}^H = 0$ are the dihedral and the tetrahedral subgroups of I (see 2.5 (4)). Only D_6, D_{10}, and T occur as isotropy groups (see 2.5 (4)), and these groups are their own normalizers (see 2.4 (1)). This implies 3.4 (3).

Let H be D_6, D_{10}, or T. Then $\Sigma^H = \{x, y\}$ consists of two points. Let x be the I fixed point. Denote the points in Σ_{4k}^H by $(z_1, \ldots, z_{4k})$ where $z_i = x$ or $z_i = y$. Then $\Sigma_{4k}^I = \{(x, \ldots, x)\}$. Let A_+ be the set of points $(z_1, \ldots, z_{4k})$ for which $z_1 = x$, but one of the z_i's is y. Let A_- be the set of points $(z_1, \ldots, z_{4k})$ such that $z_1 = y$ and $z_{4k} = x$. Let $A_0 = \{(y, \ldots, y)\}$. With these definitions 3.4 (4)–(6) can be easily checked using Lemma 2.8.

We check the assumptions in Theorem 3.5. The bundle isomorphism b from above induces an I-orientation on $T\Sigma_{4k}$ by assuming that $\mathbf{R}$ and Ω are I-oriented. Because I acts orientation preservingly on Σ and $\dim \Sigma = 3$ it follows from the definition that $\operatorname{sign}(I, \Sigma)$ vanishes. The product formula (see 3.2 (3)) implies that $\operatorname{sign}(I, \Sigma_{4k}) = 0$. We check that $\operatorname{sign}(H, \Sigma_{4k}^H) = 0$ for $W(H) = C_m$ and $m = 2$, 3, and 5. This follows because Σ^H bounds as $W(H) = C_2$-oriented manifold, and so does $\Sigma_{4k}^H = (\Sigma^H)^{4k}$. The vanishing of the the equivariant signature follows in this case from 3.2 (1) because Σ_{4k}^H is cobordant to the empty set. This completes the check of the assumptions for Theorem 3.5 and the proof of Theorem 3.1. ∎

Morimoto pointed out that the surgeries used in this paper are not possible if one uses the bundle trivialization $\varphi \times \cdots \times \varphi$ instead of the stable bundle isomorphism b which we use. This is due to a subtle relation between the framing and the assumptions in Hypothesis 3.4 which allow us to cancel isolated H fixed points for the dihedral subgroups of I.

Theorem 3.5 is proved via equivariant surgery. To discuss the relevant results we need to introduce the concept of a G-normal map. In this discussion G denotes a finite group. We give a definition which suffices in our context.

Definition 3.6. *A normal map consists of two closed smooth G-oriented manifolds X and Y, an equivariant map $f : X \to Y$, a G-vector bundle ξ over Y, a representation Γ of G, and a (stable) G-vector bundle isomorphism $b: TX \oplus \underline{\Gamma} \to f^*(\xi)$. We assume that X and Y satisfy the* *assumptions in 3.3. In addition, for any subgroup H of G this data must satisfy:*

(1) $\dim X^H = \dim Y^H$.
(2) *The degree of f^H is 1 whenever $\dim X^H \geq 1$.*
(3) *Y^H is connected and simply-connected, and X^H is connected whenever $\dim X^H \geq 1$.*

Usually we will abbreviate the data of the normal map and write (X, f, b). If we need to emphasize the target space we also write $(X, f : X \to Y, b)$.

The bundle data in this definition is considered stably. Let Γ_0 be a representation of G. The bundle isomorphism $b' = b \oplus \mathrm{Id}: TX \oplus (\underline{\Gamma} \oplus \underline{\Gamma}_0) \to f^*(\xi \oplus \underline{\Gamma}_0)$ is considered to be equivalent to b. Equivalence classes of such bundle isomorphisms are called stable G-vector bundle isomorphisms.

For normal maps $\mathcal{W} = (X, f, b)$ we have a concept of cobordism. It is called G-normal cobordism. Let $\mathcal{W}' = (X', f', b')$ be another G-normal map, and suppose both normal maps have the same target space Y. Let W be a G-oriented cobordism between X and X'. Let $F : W \to Y \times I$ be an equivariant map such that F restricts to f and f', so $f = F_{|X} : X \to Y \times 0 = Y$ and $f' = F'_{|X'} : X' \to Y \times 1 = Y$. Let $B: TW \oplus \underline{\Gamma} \to F^*(\xi \times I)$ be a stable G-vector bundle isomorphism which restricts to b over X and to b' over X'. Then (W, F, B) is called a G-normal cobordism between $\mathcal{W}$ and $\mathcal{W}'$. It is relative to the L fixed point set, $L \subset G$, if all data restrict over the L fixed point set to a product with the unit interval.

In the process of surgery we need to refer to Wall's surgery obstruction groups. The groups $L^s_k(\mathbf{Z}[G], w)$ and $L^h_k(\mathbf{Z}[G], w)$ are defined in [**W1**] and [**W2**]. We summarize the properties and computations needed in this paper. Set

(3.7) $$L^h(G) := L^h_0(\mathbf{Z}[G], 1) \quad \text{and} \quad L^s(G) := L^s_0(\mathbf{Z}[G], 1)$$

The groups are related to each other by the exact Rothenberg sequence [**Sh**]

(3.8) $$H^1(\mathbf{Z}_2, \mathrm{Wh}(G)) \to L^s(G) \to L^h(G) \xrightarrow{\alpha} H^0(\mathbf{Z}_2, \mathrm{Wh}(G))$$

where the first and last terms are Tate cohomology groups of $\mathbf{Z}_2$ with coefficients in the Whitehead group $\mathrm{Wh}(G)$ of G. There exist signature homomorphisms

(3.9) $$\mathrm{sign}: L^s(G) \to R(G) \quad \text{and} \quad \mathrm{sign}: L^h(G) \to R(G)$$

The signature homomorphisms commute with the forgetful map $L^s(G) \to L^h(G)$. The Wall groups are abelian and

Proposition 3.10 (see [**W1**, p. 167]). *The kernels of the signature homomorphisms are finite. If the surgery obstruction group is torsion free, then the obstruction is detected by the signature.*

Theorem 3.11. *The Wall groups $L^h(C_2)$, $L^h(D_4)$, $L^h(D_6)$, and $L^s(D_{10})$ are torsion-free.*

PROOF: Wall [**W1**, p. 162] computed that $L^s(C_2) \cong \mathbf{Z} \oplus \mathbf{Z}$, and $L^s(C_2) \cong L^h(C_2)$ because $\mathrm{Wh}(C_2)$ is trivial. For the computation of $L^h(D_4)$ and $L^s(D_4)$ see [**Ba**, p. 388] and [**W2**, p. 72]. Wall showed [**W2**, p. 72 and p. 74] that $L^s(D_6)$ and $L^s(D_{10})$ are torsion free. But, $\mathrm{Wh}(D_6)$ vanishes (see [**O**]), and this implies that $L^h(D_6)$ is torsion free. ∎

A group H is called 2-hyperelementary if there is a short exact sequence

$$1 \to K \to H \to P \to 1$$

where K is cyclic of odd order, and P is a two-group. There are restriction maps Res: $L^h(G) \to L^h(H)$ whenever H is a subgroup of G. Dress showed [**Dr**]

Theorem 3.12. *The sum of the restriction maps*

$$\mathrm{Res} : L^h(G) \to \bigoplus_{H \in \mathcal{H}} L^h(H)$$

is injective if $\mathcal{H}$ is the set of all (maximal) 2-hyperelementary subgroups of G.

We will also use the following result on the reduced projective class group.

Proposition 3.13 [**RU**]. *Every finitely generated projective $\mathbf{Z}[H]$ module is stably $\mathbf{Z}[H]$ free if $H = C_2$ or $H = A_5$.*

Our next theorem and the proposition following it will be important tools in the proof of Theorem 3.5.

Theorem 3.14. *Suppose $(X, f : X \to Y, b)$ is a G-normal map, $H \subset G$ is an isotropy group of the action on X, and*

(1) *f^K is a homotopy equivalence whenever $H \subset K \subset N_I H$ and K/H is a non-trivial group of prime power order.*
(2) *All finitely generated projective $\mathbf{Z}[W(H)]$ modules are $\mathbf{Z}[W(H)]$ stably free.*

Then there exists an element $\sigma_H(f,b) \in L^h(W(H))$ with the following properties

(a) $\text{sign}(\sigma_H(f,b)) = \text{sign}(W(H), X^H) - \text{sign}(W(H), Y^H)$.

(b) *If $\sigma_H(f,b) = 0$, then (X,f,b) is G-normally cobordant to a normal map (X_0, f_0, b_0) such that f_0^H is a homotopy equivalence.*

(c) *The normal cobordism in (b) may be taken relative to all L fixed point sets, such that L is not conjugate to a subgroup of H.*

This theorem is a standard result in equivariant surgery theory (see [**DR1**], [**PR2**], or [**D**, Section 3]). The essential conclusion of Hypothesis 3.4 is as follows. The proof is easy for specialists, but somewhat too technical to be carried out here.

Proposition 3.15. *Let X be a smooth G-oriented manifold which satisfies Hypothesis 3.4. There exists a smooth G-oriented manifold X_1 with these properties*

(1) *X and X_1 are cobordant as G-oriented manifolds.*

(2) *X_1^H consists of exactly two points if H is a proper subgroup of G and $\dim X^H = 0$.*

(3) *There exists a G-vector bundle isomorphism $b_1 : \underline{\mathbf{R}}^m \oplus TX_1 \to \underline{\mathbf{R}}^m \oplus \Omega$ for some m and Ω as in 3.4.*

(4) *If $K \subset G$, X^K is connected and of dimension ≥ 2, then X_1^K is also connected.*

(5) *The cobordism is relative to the G fixed point set.*

One important computation of part of a surgery obstruction is based on our next proposition. There we consider an H-normal map $\mathcal{W} = (X, f : X \to Y, b)$ such that the associated H-vector bundle ξ over Y is a product bundle $Y \times \Upsilon$ (see 3.6). Here Υ is a representation of H. Assume that $Y \times \Upsilon \cong TY \oplus \underline{\Gamma}$, and the bundle isomorphism is of the form $b : TX \oplus \underline{\Gamma} \to f^*(Y \times \Upsilon)$. This implies that $T_xX \cong T_{f(x)}Y$ as a representation of H_x for all $x \in X$. With this bundle data we have

Proposition 3.16. *Suppose for every proper subgroup K of H the induced map f^K is a homotopy equivalence. There is an element $\sigma_1(f,b) \in L^h(H)$ which has the same properties as $\sigma_1(f,b)$ in 3.14. Let $\alpha(\sigma_1(f,b))$ be its image in $H^0(\mathbf{Z}_2, \text{Wh}(H))$ (see 3.8). Then $\alpha(\sigma_1(f,b)) = 0$.*

PROOF OF THEOREM 3.5: Throughout the proof all cobordisms will be relative to the I fixed point set. It follows from Proposition 3.15 that there is a G-oriented manifold X_1 which is G-cobordant to X as G-oriented manifold such that

(1) $X_1^H = S^0$ for $H = T$ and $H = D_{2m}$ for $m = 2$, 3, and 5.

(2) There exists a G-vector bundle isomorphism $b_1 : \underline{\mathbf{R}}^m \oplus TX_1 \to \underline{\mathbf{R}}^m \oplus \underline{\Omega}$ for some m and Ω as in 3.4.
(3) X_1^K is connected whenever $\dim X_1^K > 0$.
(4) X_1^I consists of exactly one point (we denote it by x).

To derive this conclusion we use that X satisfies Hypothesis 3.4.

We use (X_1, b_1) to construct a normal map. Note that $\Omega = T_x X_1$. Use the Thom Pontrjagin collapse (see the description after 2.6) to construct a map $f_1: X_1 \to Y := S(\Omega \oplus \mathbf{R})$. Here we map $0 \in T_x X_1$ to $(0,1) \in S(\Omega \oplus \mathbf{R})$ and use Ω as $\bar{I}$-oriented tangent representation at $(0,1)$ to give an I-orientation on Y. As bundle Ξ choose $\underline{\Omega} \oplus \underline{\mathbf{R}}$ and let b_1 be as in (2). Obviously, (X_1, f_1, b_1) is an I normal map (for the definition see 3.6).

Let $H = C_m$ for $m = 2$, 3, or 5. In either case, $W(H) = NH/H$ is C_2, and the assumptions of 3.14 are satisfied for these H. Consider the obstruction $\sigma_H(f_1, b_1) \in L^h(C_2)$ defined by 3.14. By assumption, $\mathrm{sign}(W(H), X^H) = 0$. It follows from the bordism invariance of the signature (see 3.2 (1)) that $\mathrm{sign}(W(H), X_1^H) = 0$. It follows from 3.2 (2) for the sphere Y^H that $\mathrm{sign}(W(H), Y^H) = 0$. Then $\mathrm{sign}(\sigma_H(f_1, b_1)) = 0$ by 3.14 (a). It follows from 3.10 and 3.11 that $\sigma_H(f_1, b_1) = 0$. By 3.14 (b), (X_1, f_1, b_1) is I-normally cobordant to a normal map (X_2, f_2, b_2) such that f_2^H is a homotopy equivalence. We apply this procedure for $m = 2$, 3, and 5. Then we get an I-normal map (X_3, f_3, b_3) such that f^H is a homotopy equivalence for $H = C_m$ and D_{2m} for $m = 2$, 3, and 5. The case where H is a dihedral group was obtained in the first step, and in the second step we used cobordisms which are relative to the D_{2m} fixed point sets (see 3.14 (c)).

Due to our construction (X_3, f_3, b_3) satisfies the assumptions of 3.14 for $H = 1$. We apply once more 3.14 to define $\sigma_1(f_3, b_3) \in L^h(I)$. By Dress' Induction Theorem (see 3.12) we need to show that $\mathrm{Res}_{D_{2m}}(\sigma_1(f_3, b_3)) = 0$ to conclude that $\sigma_1(f_3, b_3) = 0$. The restricted surgery obstruction is obtained by considering the normal map with the restricted group action of D_{2m}. So

$$\mathrm{Res}_{D_{2m}}(\sigma_1(f_3, b_3)) = \sigma_1(\mathrm{Res}_{D_{2m}}(f_3, b_3))$$

For $m = 2$ and 3, $\mathrm{Res}_{D_{2m}}(\sigma_1(f_3, b_3))$ lies in a torsion-free group (see 3.11), it is detected by signatures (see 3.11), and it vanishes due to the bordism invariance of the signature (see 3.2 (1)) and the assumption that $\mathrm{sign}(D_{2m}, X) = \mathrm{sign}(D_{2m}, Y) = 0$.

We analyze $\mathrm{Res}_{D_{10}}(\sigma_1(f_3, b_3))$ which we abbreviate as σ. The assumptions of Proposition 3.16 apply to $\mathrm{Res}_{D_{10}}(X_3, f_3, b_3)$. The conclusion of Proposition 3.16 says that $\alpha(\sigma) = 0$. It follows from the exact Rothenberg sequence (3.8) that σ is the image of some element $\sigma' \in L^s(D_{10})$. Again,

σ' is an element in a torsion-free group (3.11) and detected by signatures (3.10). Then $\text{sign}(D_{10}, X_3) - \text{sign}(D_{10}, Y) = 0$ by 3.14 (a) and $\sigma' = 0$ by arguments as above. We showed that $\text{Res}_H(\sigma_1((f_3, b_3))$ vanishes for $H = D_{2m}$ and $m = 2$, 3, and 5. Thus $\sigma_1(f_3, b_3) = 0$ by 3.12. We apply once more 3.14 (b) and conclude that (X_3, f_3, b_3) is I-normally cobordant to a normal map (X_0, f_0, b_0) whose underlying function f_0 is a homotopy equivalence. The manifold X_0 has all of the properties required in the theorem, and this concludes the proof. ∎

The final proof of this section is included for reasons of completeness. It involves some advanced notions from transformation groups which the non-specialist may not be familiar with.

PROOF OF PROPOSITION 3.16: Let f^s denote the restriction of f to the non-free orbits. It has been shown in [**D**, Theorem B] that

$$\textbf{(3.17)} \qquad \alpha(\sigma(f,b)) = [(1+T)\tau(f^s) - T\,\text{tr}_{S(E)}\,\tau(f^s) + T\tau(\phi_{|Y^s})]$$

Here [ ] stands for the class in $H^0(\mathbf{Z}_2, \text{Wh}(H))$, τ is the generalized Whitehead torsion, T is the natural conjugation on the generalized Whitehead group $\widetilde{\text{Wh}}(H)$, $\text{tr}_{S(E)}$ is the transfer for the generalized Whitehead group with respect to the sphere bundle $S(E)$, and E and ϕ are an H-vector bundle and an H-fibre homotopy equivalence associated with the bundle isomorphism b. All of these terms are explained in detail in [**D**] and [**DR2**]. We compute the right hand side of Formula 3.17.

The generalized Whitehead group $\widetilde{\text{Wh}}(H)$ decomposes into a direct sum $\bigoplus \text{Wh}(W(K))$ where K ranges over subgroups of H, one in each conjugacy class of subgroups of H. Observe that $\tau(f^s)$ does not have a component in the top summand $\text{Wh}(H)$ of the direct sum decomposition because $f^s : X^s \to Y^s$ involves only non-free orbits. Also, T leaves the summands invariant. Thus $(1+T)\tau(f^s)$ does not have a component in $\text{Wh}(H)$. By definition, E is the bundle which pulls back to the tangent bundle of X. The transfer may be computed stably, and in our situation we may use the product bundle $E = \underline{\Upsilon}$. This means that $\text{tr}_{S(E)}\,\tau(f^s) = [S(\Upsilon)]\tau(f^s)$. Here $[S(\Upsilon)]$ denotes the class of $S(\Upsilon)$ in the Burnside ring. The product is defined in [**DR2**]. Next we use that all fixed points sets Y^H, $H \subset G$, have the same dimension modulo two, and the product formula in [**DR2**, Proposition 0.1]. We conclude that $\text{tr}_{S(E)}\,\tau(f^s)$ is an integral multiple of $\tau(f^s)$. So it does not have a component in the summand $\text{Wh}(H)$ of $\widetilde{\text{Wh}}(H)$.

Finally, we consider the fibre homotopy equivalence ϕ. We assumed that b is a stable trivialization, and this implies that $\phi_{|Y^s} : Y^s \times \Upsilon \to Y^s \times \Upsilon \cong$

$TY^s \oplus \underline{\Gamma}$ may be taken as a product, $\phi_{|Y^s} = \text{Id} \times \phi_x$. The map in the fibre direction is (homotopic) to the identity, and the torsion of ϕ_x vanishes. By the product formula [**DR2**, Proposition 0.1] $\tau(\phi_{|Y^s}) = [Y^s]\tau(\phi_x) = 0$.

This concludes the computation of the terms which occur on the right hand side in the Formula 3.17. None of them contributes in $\text{Wh}(H)$, and $\alpha(\sigma(f,b)) = 0$. This concludes the proof of the proposition. ∎

4. Remarks on Real Algebraic *G*-varieties

This section is preparation for the proof of Theorem 1.3. Throughout, G will denote a compact Lie group. All representations of G will be orthogonal. Let Ξ and Ω be representations of G. Then $C^r(\Xi, \Omega)$ denotes the set of all C^r maps from Ξ to Ω. On this space we can consider the C^ρ topology for $\rho \leq r$ [**H**]. A polynomial map $f : \Xi \to \Omega$ is a map of the form $f = (f_1, \ldots, f_m)$ where each coordinate f_i of f is a polynomial.

The averaging operator A will play a role. We discuss it first. Let $f : \Xi \to \Omega$ be a map between representations of G. Denote the Haar measure of G by dg, and let x be any point in Ξ. Then

$$A(f)(x) = \int_G g^{-1} f(gx)\, dg.$$

We collect some standard facts concerning the averaging operator A.

Lemma 4.1.

(1) *$A(f)$ is equivariant, and $A(f) = f$ if f is equivariant.*
(2) *If $0 \leq r \leq \infty$ and $f \in C^r(\Xi, \Omega)$, then $A(f) \in C^r(\Xi, \Omega)$.*
(3) *If f is a polynomial map then so is $A(f)$.*
(4) *The selfmap of $C^r(\Xi, \Omega)$ induced by A (see (2)) is continuous with respect to the C^ρ topology for any $\rho \leq r$. We also denote this map by A. We will need the case $r = \infty$ and $\rho = 1$.*
(5) *If $f \in C^1(\Xi, \Omega)$, then $|D^s A(f(x))| \leq |D^s f(x)|$ for all partial derivatives D^s of order at most one.*

PROOF: (1) follows immediately from the definition. To see (2) and (3) observe that

$$D^s(A(f))(x) = \int_G D^s(g^{-1} f g)(x)\, dg$$

for all $s = (s_1, \ldots, s_n)$. Here g stands also for the map induced by $g \in G$, and $D^s = \partial^{|s|}/\partial x_1^{s_1} \ldots \partial x_n^{s_n}$, where $|s| = s_1 + \cdots + s_n$. (2) follows immediately from this. (3) follows because a function is a polynomial map if all of its derivatives eventually vanish. (4) Let $p_{K,q}$ be the semi-norm on the set $C^r(\Xi, \Omega)$ defined by

$$p_{K,q}(f) = \sup_{|s| \leq q, x \in K} |D^s f(x)|$$

where K is a compact subset of Ξ, $q \leq r$ and $|y| = \sqrt{y_1^2 + \cdots + y_m^2}$ for $y = (y_1, \ldots, y_m) \in \Omega$. For $q \leq \rho \leq r$ these semi-norms induce the C^ρ-topology on $C^r(\Xi, \Omega)$. Let $K_j \subset \Xi$ be the disk of radius j. These disks are G-invariant sets. The family $p_{K_j,q}$ also induces the C^ρ-topology on $C^r(\Xi, \Omega)$.

It follows from the formula for the higher partial derivatives of a composition that:

$$\text{If} \quad p_{K_j,q}(f) < \epsilon \quad \text{then} \quad p_{K_j,q}(g^{-1} f g) < C_q \epsilon.$$

for some appropriate constant C_q which does not depend on K. One finds:

$$\text{If} \quad p_{K_j,q}(f) < \epsilon \quad \text{then} \quad p_{K_j,q}(A(f)) < C_q \epsilon.$$

This implies that A is continuous. (5) This follows from the observation that $C_0 = C_1 = 1$ in the proof of (4). ∎

Definition 4.2. *A real algebraic G-variety is a G-invariant subset V of some G-representation Ξ which has the form*

$$V = \{x \in \Xi \mid p(x) = 0 \text{ for all } p \in I\}$$

where I is a set of polynomial functions from $\Xi \cong \mathbf{R}^n$ to $\mathbf{R}$. We denote by $\mathcal{I}(V)$ the ideal of polynomials p such that $p(V) = 0$.

CONVENTION: Until now we were very explicit in our notation and emphasized throughout that the varieties which we consider are real algebraic. In the following we will omit this adjective, but we assume that it is understood.

Definition 4.3 (see [**AK1**]). *A point x in an algebraic G-variety $V \subset \Xi$ is called non-singular of dimension d in V if there are polynomials $p_i \in \mathcal{I}(V)$, $i = 1, \ldots, k-d$ ($k = \dim \Xi$), and a neighborhood U of x in Ξ such that*

(1) $V \cap U = U \cap \bigcap_{i=1}^{k-d} p_i^{-1}(0)$,
(2) *the gradients $(\nabla p_i)_x$, $i = 1, \ldots, k-d$ are linearly independent.*

Definition 4.4. *For an algebraic G-variety V,*

(1) $\dim V = \max\{d \mid$ *there is an* $x \in V$ *which is non-singular of dimension* $d\}$,
(2) $\operatorname{Nonsing} V = \{x \in V \mid x$ *is non-singular of dimension* $\dim V\}$,
(3) $\operatorname{Sing} V = V \setminus \operatorname{Nonsing} V$.

Facts 4.5.

(1) $\operatorname{Sing} V$ *is an algebraic G-variety.*
(2) $\dim \operatorname{Sing} V < \dim V$.
(3) $\operatorname{Nonsing} V$ *is a smooth G-manifold.*
(4) *The finite union, intersection, and cartesian product of algebraic G-varieties are algebraic G-varieties.*

Definition 4.6. *Let Ξ and Ω be representations of G. Let $V \subset \Xi$ and $W \subset \Omega$ be algebraic G-varieties and let X be an algebraic G-subvariety of V. A G-map $f: V - X \to W$ is called a G-equivariant rational function (or if G is understood, an equivariant rational function) if there are equivariant polynomials $p: V \to \Omega$ and $q: V \to \mathbf{R}$ such that $q^{-1}(0) \subset X$ and $f(x) = p(x)/q(x)$ for all $x \in V - X$. If X is empty we say f is an equivariant entire rational function.*

Lemma 4.7. *Any compact algebraic G-variety $V \subset \Xi$ is the set of zeros of a proper G-invariant polynomial (proper means that inverse images of compacta are compact).*

PROOF: First we show that there is a G-invariant polynomial p such that $V = p^{-1}(0)$. The proof is as follows. The ideal $\mathcal{I}(V)$ is finitely generated. Let $p_1, \ldots, p_k$ be generators, and set $p_V = p_1^2 + \cdots + p_k^2$. The average $A(p_V)$ is a G-invariant polynomial by Lemma 4.1 (3). Since $p_V^{-1}(0) = V$ and p_V is non-negative, $A(p_V)^{-1}(0) = V$. Next we need to modify $p = A(p_V)$ such that it is also proper.

Take any G-invariant polynomial p with $p^{-1}(0) = V$, and remember that V is compact. It is the statement of Lemma 1.7 in [**AK1**] there are a positive number c, a number d and a neighborhood N of V such that

$$|p(x)| \geq c|x|^d \quad \text{for all } x \in \Xi - N.$$

The polynomial $(1 + |x|^2)^{|d|+1} p(x)$ has all of the properties asked for in the lemma.

5. Approximating G-manifolds by G-varieties

We follow the argument of Akbulut and King in [**AK1**, Section 2] and generalize it to the equivariant situation. In the non-equivariant situation they prove a stronger result than our Theorem 1.3, and for this reason we may omit some of their considerations.

The approximations required in this section are C^1-approximations of C^∞-functions. This will be understood whenever we use the word approximation. Sometimes we require approximations which hold not only on a compact set but everywhere on the underlying space of a representation (see 5.1). We will be explicit if such an approximation is used. There are two more notions from differential topology which play an essential role. These are the concepts of transversality and of isotopy. We discuss them.

Let $f : M \to N$ be a smooth map between smooth manifolds. Let Y be a smooth submanifold of N. Then f is said to be *transverse* to Y if for all $y \in Y$ and for all $x \in f^{-1}(y)$ the composition

$$T_xM \xrightarrow{Df_x} T_yN \to \nu_y(Y,N)$$

is surjective. Here $\nu_y(Y,N)$ stands for the normal space to Y in N at y. The second map is the projection on the second factor in the direct sum decomposition $T_yN = T_yY \oplus \nu_y(Y,N)$. Observe that $f^{-1}(Y)$ is a smooth submanifold of M if f is transverse to Y. Transversality is an open condition. That means, if $f : M \to N$ is transverse to $Y \subset N$, then there is a neighborhood U of f in $C^\infty(M,N)$ with the C^1-topology such that every function $f' \in U$ is also transverse to Y. This applies if M is compact. In case of a non-compact manifold M we will have to apply some additional arguments to control the situation outside of a compact set.

Let X be a smooth submanifold of M, and let j be the inclusion. An *isotopy* of X is a smooth map $h : X \times I \to M$ such that $h(x,0) = j(x)$ and $h(\ ,t) : X \times \{t\} \to M$ is an embedding for each $t \in I$. With the obvious identifications this last map is also denoted by $h_t : X \to M$. The isotopy is said to be small if h_t is close to j for each $t \in I$. An ambient isotopy is an isotopy of M. Two submanifolds X and X_1 of M are called isotopic if there is an isotopy $h : X \times I \to M$ such that $h(X,1) = X_1$, and h is called an isotopy between X and X_1. The isotopy fixes $L \subset X$ if $h(x,t) = j(x)$ for all $x \in L$ and for all $t \in I$. To define equivariant isotopies and equivariant ambient isotopies we assume that all spaces have a G-action and that all maps are equivariant.

Let $H : M \times I \to N$ be a smooth map. It restricts to maps H_t: $M \times \{t\} = M \to N$, $t \in I$. Suppose H_t is transverse to $Y \subset N$ for each $t \in I$. It is a theorem that $H_0^{-1}(Y)$ and $H_1^{-1}(Y)$ are isotopic. This theorem follows from the Morse Lemma. The homotopy H between H_0 and H_1 is called *small* if H_t and H_0 are close to each other for all $t \in I$. If the homotopy is small, then the isotopy between $H_0^{-1}(Y)$ and $H_1^{-1}(Y)$ may be chosen small as well. This one may check easily. If the homotopy is relative to $L \subset M$ (that is, $H(x,t) = H(x,0)$ for all $x \in L$ and $t \in I$) then the isotopy may be chosen to fix L. The connection between transversality and isotopy explained in this paragraph will be used repeatedly in this section. We refer to it as a *transversality argument.*

The isotopy extension theorem provides extensions of an isotopy h of a submanifold X of M to an ambient isotopy of M which extends h. This discussion also applies in the equivariant setting.

Lemma 5.1 (Compare [**AK1**, Lemma 2.1]). *Let L be a compact non-singular algebraic G-variety in Ξ and let $f: (\Xi, L) \to (\mathbf{R}, 0)$ be a G-invariant smooth function. Suppose f has compact support. Then there are equivariant entire rational functions $u: (\Xi, L) \to (\mathbf{R}, 0)$ which approximate f arbitrarily closely on all of Ξ. More precisely, given $\epsilon > 0$ we can find an equivariant entire rational function $u: (\Xi, L) \to (\mathbf{R}, 0)$ such that $|D^s(f-u)(x)| < \epsilon$ for all $x \in \Xi$ and all partial derivatives D^s with $|s| \leq 1$.*

PROOF: Let $S \subset \mathbf{R} \times \Xi$ be the unit sphere and let $\theta: S - (1,0) \to \Xi$ be the stereographic projection $\theta(t,x) = x/(1-t)$. We define the function $h: S \to \mathbf{R}$ by

$$h(x) = \begin{cases} f(\theta(x)) & \text{for } x \in S - (1,0) \\ 0 & \text{for } x = (1,0). \end{cases}$$

Because f is a smooth function with compact support, h is smooth. Because θ is rational it follows that $(1,0) \cup \theta^{-1}(L)$ is an algebraic G-variety (see [**AK1**, Lemma 1.3]); in fact, $\theta^{-1}(L)$ is non-singular. Also, $h(x) = 0$ for $x \in S$ near $(1,0)$. By the first part of Lemma 2.1 of [**AK1**] we may approximate h closely by a (non-equivariant) polynomial $v : (S^n, (1,0) \cup \theta^{-1}(L)) \to (\mathbf{R}, 0)$. By our Lemma 4.1 (4), $A(v)$ closely approximates $A(h)$ on S because S is compact. By Lemma 4.1 (1) $A(h) = h$ because θ and f, and hence also h are equivariant. Given any $\epsilon > 0$ we may choose v close enough to h such that

$$|D^s(f - A(v) \circ \theta^{-1})(x)| < \epsilon$$

for all $x \in \Xi$ and all partial derivatives D^s with $|s| \leq 1$. By Lemma 4.1 (3), $A(v)$ is a G-invariant polynomial. Set $u = A(v) \circ \theta^{-1}$. This function u is an equivariant entire rational function because $\theta^{-1}(x) = (|x|^2 - 1, 2x)/(|x|^2 + 1)$ is an equivariant entire rational function, and u approximates f in the way stated in the lemma. ∎

Definition. *Let Z be a topological G-space and Y a G-invariant subset of Z. We say that Y separates Z compactly and equivariantly if there are closed G-invariant sets Z_0 and Z_1 such that $Z = Z_0 \cup Z_1$, $Y = Z_0 \cap Z_1$ and Z_1 is compact.*

Lemma 5.2 (Compare [**AK1**, Lemma 2.2]). *Let L and Z be algebraic G-varieties and let Y be a closed, codimension 1 G-submanifold of* Nonsing Z *which contains L and separates Z compactly and equivariantly. Suppose L is non-singular. Then there are arbitrarily small G-isotopies of* Nonsing Z *which fix L and carry Y to a non-singular algebraic G-variety.*

PROOF: Let Ξ be a representation of G in which the algebraic G-varieties L and Z are realized as zero sets of some polynomials. By Lemma 4.7 there is a proper G-invariant polynomial $q\colon \Xi \to \mathbf{R}$ such that $q^{-1}(0) = L$. Choose G-invariant sets Z_0 and Z_1 such that $Z = Z_0 \cup Z_1$, $Y = Z_0 \cap Z_1$ and Z_1 is compact. We want to choose a smooth function $f : Z \to \mathbf{R}$ such that

(1) $f_{|Z_0}$ is non-negative, $f_{|Z_1}$ is non-positive, and $f^{-1}(0) = Y$,
(2) $f(x) = q^2(x)$ for all x outside some G-invariant compact subset of Z,
(3) 0 is a regular value of f,
(4) f is equivariant.

At this point Akbulut and King choose a (possibly) non-equivariant function f_0 which has Properties (1)–(3). Let $f = A(f_0)$ be the average. This function obviously has properties (1), (2), and (4). We check that it also satisfies (3). That f_0 has 0 as regular value means that the derivative of f_0 is strictly positive (or negative) as we go from Y in the normal direction in Z_0 (or Z_1). This property is preserved when we form the average $f = A(f_0)$.

Extend f to a G-invariant smooth function $\tilde{f}$ on Ξ which agrees with q^2 outside some G-invariant compact subset of Ξ. Approximate $\tilde{f} - q^2 : \Xi \to \mathbf{R}$ in the way described in Lemma 5.1 by an equivariant entire rational function $u : (\Xi, L) \to (\mathbf{R}, 0)$. Set $v = u + q^2$. For a given $\epsilon > 0$ we may suppose that $|D^s(\tilde{f} - v)(x)| < \epsilon$ for any $x \in \Xi$ and for any s of order at most 1. Let $h : \Xi \times I \to \mathbf{R}$ be the homotopy between $\tilde{f}$ and v defined by $h_t = t\tilde{f} + (1-t)v$. It is smooth and equivariant. Furthermore, $\tilde{f}$ and h are proper. We consider h_t as a function from Ξ to $\mathbf{R}$. As we approximated a proper function one concludes for a sufficiently small ϵ that $(h_{t|Z})^{-1}(0) = h_t^{-1}(0) \cap Z$ is near $f^{-1}(0) = Y$ for any $t \in I$. As we used C^1 approximations, it follows that each $h_{t|Z} : Z \to \mathbf{R}$ is transverse to zero on a preassigned bounded set. Also, no point outside of this bounded set will map to zero because h is proper. It follows from the transversality argument in the beginning of this section that there is an ambient isotopy of Nonsing Z which carries $Y = f^{-1}(0)$ to $v^{-1}(0) \cap Z$. Because we assumed that u vanishes on L, it follows that $h_t = f$ on L for all $t \in I$. For the isotopy it means that it fixes L. Since $Y \subset$ Nonsing Z and $v_{|Z}$ approximates f, we may assume $v^{-1}(0) \cap Z \subset$ Nonsing Z. Hence $v^{-1}(0) \cap Z$ is a non-singular algebraic variety by [**AK1**, Lemmas 1.3 and 1.4], so it is a non-singular algebraic G-variety because it is G-invariant. This proves the lemma. ∎

Let Ξ be an orthogonal representation of G with underlying vector space $\mathbf{R}^n$. The action of G on $\mathbf{R}^n$ induces via conjugation an action of G on the set of $n \times n$ matrices $M_n(\mathbf{R}) = \mathbf{R}^{n^2}$. We denote this representation by $H(\Xi)$. In $H(\Xi)$ we have the non-singular G-variety

$$G(\Xi, k) = \{L \in \mathbf{R}^{n^2} \mid L^2 = L,\ L \text{ is symmetric, and } \operatorname{Trace} L = k\}.$$

Given a subspace U of $\mathbf{R}^n$ and an element $g \in G$ we let it act on U mapping (g, U) to gU. This defines an action of G on the Grassmannian of k-dimensional subspaces in Ξ. The identification of a k-plane with the matrix of orthogonal projection onto the plane gives an equivariant diffeomorphism from $G(\Xi, k)$ to the Grassmannian of k-dimensional subspaces in Ξ. Similarly, the total space of the universal k-plane bundle

$$E(\Xi, k) = \{(L, y) \in \mathbf{R}^{n^2+n} \mid L \in G(n, k),\ Ly = y\}$$

is a G-variety in $H(\Xi) \times \Xi$.

Lemma 5.3 (Compare [**AK1**, Lemma 2.3]). *Let $V \subset \Xi$ be a non-singular algebraic G-variety. Then there is an equivariant entire rational function $\eta: (\Xi, V) \to (H(\Xi), G(\Xi, k))$ such that for each $x \in V$, $\eta(x)$ is orthogonal projection onto the normal k-plane to V at x ($k = \dim \Xi - \dim V$).*

PROOF: The property that $\eta(x)$ is orthogonal projection onto the normal k-plane to V at x defines the restricted function $\eta_{|V} : V \to G(\Xi, k)$. Obviously $\eta_{|V}$ is equivariant because if $x \in V$ and N is the normal plane at x, then gN is the normal plane at gx for any $g \in G$. Akbulut and King give explicit formulae for polynomials $p_0: \Xi \to H(\Xi)$ and $q_0: \Xi \to \mathbf{R}$ such that $\eta_{|V} = p_0/q_0$ on V. Let $p = A(p_0)$ and $q = A(q_0)$ be their averages. Because $\eta_{|V}$ is equivariant it is immediate that $p/q = \eta_{|V}$ on V.

The polynomial q is nowhere zero on V, and by the construction of Akbulut and King, $q(x)$ is non-negative. Since V is an algebraic G-variety, there is a G-invariant non-negative polynomial q_V on Ξ such that $q_V^{-1}(0) =$

V (see the proof of Lemma 4.7). The sum $q + q_V$ is nowhere zero on Ξ and agrees with q on V. This shows that $\eta = p/(q+q_V) : \Xi \to H(\Xi)$ is an equivariant entire rational function with all of the desired properties. ∎

Lemma 5.4 (Compare [**AK1**, Lemma 2.4]). *Let $L \subset \Xi$ and $W \subset \Omega$ be non-singular algebraic G-varieties. Let $T \subset \Xi$ be a G-invariant compact set containing L and let $\gamma : T \to W$ be a smooth G-map. Suppose there is an equivariant entire rational function $u: (\Xi, L) \to (\Omega, W)$ such that $\gamma_{|L} = u_{|L}$.*

Then there are an algebraic G-variety $U \subset \Xi \times \Omega$, a G-invariant subset $J \subset U \cap (T \times \Omega)$, an equivariant entire rational function $w: U \to W$ and a smooth G-map $\varphi : (T, L) \to (\Omega, 0)$ such that

(1) $J = \{(x, \varphi(x)) \mid x \in T\}$,
(2) *J is a relatively open G-invariant subset of $U \cap (T \times \Omega)$,*
(3) *The G-map $\mu : T \to W$ defined by $\mu(x) = w(x, \varphi(x))$ can be as close to γ as we wish in the C^1 topology (for some choice of U, J, w and φ),*
(4) *$\mu(x) = w(x, 0) = \gamma(x)$ for $x \in L$,*
(5) *If $x \in T$, then $(x, \varphi(x))$ is non-singular of dimension n in U where $n = \dim \Xi$,*
(6) *φ is close to zero.*

PROOF: The following construction of U, w, φ and J is as in [**AK1**]. Apply Lemma 5.3 to the non-singular G-variety W in Ω. Then we find equivariant polynomials $p: \Omega \to H(\Omega)$ and $q: \Omega \to \mathbf{R}$ such that for every x in W the value $p(x)/q(x)$ is orthogonal projection onto the normal plane to W at x. In particular, $p(x)/q(x)$ is an element in $G(\Omega, \dim \Omega - \dim W)$. We want to construct an equivariant polynomial $\psi : (\Xi, L) \to (\Omega, 0)$ which approximates $\gamma - u$ on T. A (possibly) non-equivariant polynomial $\psi_0: (\Xi, L) \to (\Omega, 0)$ of this kind exists according to the first part of [**AK1**, Lemma 2.1] because T is compact and L is non-singular. Let $\psi = A(\psi_0)$ be the average. Set $v = \psi + u$. This function $v: \Xi \to \Omega$ is an equivariant entire rational function which approximates γ on T. Define

$$U = \{(x, y) \in \Xi \times \Omega \mid v(x) + y \in W, p(v(x) + y)(y) = q(v(x) + y)y\}$$
$$w(x, y) = v(x) + y$$
$$\varphi(x) = \text{ the vector from } v(x) \text{ to the closest point to } v(x) \text{ on } W$$
$$J = \{(x, y) \in U \mid x \in T, \ y = \varphi(x)\}$$

The sets U and J are G-invariant, and w and φ are equivariant because all of the data involved is equivariant. The check of (2), (3) and (4) is easy and (5) is proved in [**AK1**, Lemma 2.4]. Finally, $\varphi(x)$ is close to zero because v approximates γ, and this implies (6). ∎

The following proposition (the case where L is empty) is an equivariant generalization of a result of Nash [**N**]. It uses the idea of a germ. Let M and M' be submanifolds of $\mathbf{R}^n$, and let L be a submanifold of M and M'. We say that the germ of M at L is the germ of M' at L if $T_xM = T_xM'$ as subspaces of $T_x\mathbf{R}^n$.

Proposition 5.5 (Compare [**AK1**, Proposition 2.8]). *Let M be a closed smooth G-submanifold of Ξ and let $L \subset \Xi$ be a non-singular algebraic G-variety contained in M. Suppose that the germ of M at L is the germ of a non-singular algebraic G-variety V at L, $L \subset V \subset \Xi$. Then there are a representation Ω of G, a G-isotopy h_t of $\Xi \times \Omega$, and an algebraic G-variety $Z \subset \Xi \times \Omega$ such that*

(1) *$h_1(M \times 0)$ is a union of non-singular components of Z*
(2) *$h_t(x,0) = (x,0)$ for all $x \in L$, $t \in [0,1]$.*
(3) *h_t can be as small as we want (in the C^1-topology on $C^\infty(\Xi \times \Omega, \Xi \times \Omega)$).*

PROOF: Let $n = \dim \Xi$ and $k = n - \dim M$. As representation Ω we use $H(\Xi) \times \Xi$. Furthermore we use the embeddings

$$E(\Xi,k) \subset G(\Xi,k) \times \Xi \subset H(\Xi) \times \Xi \overset{\text{Def}}{=} \Omega$$

As a first step we construct a small G-invariant tubular neighborhood T of M in Ξ, a smooth G-map $\gamma: T \to E(\Xi,k)$ and an equivariant entire rational function $u: (\Xi, L) \to (H(\Xi) \times \Xi, E(\Xi,k))$ such that

(a) γ is transverse to $G(\Xi,k)$
(b) $\gamma^{-1}(G(\Xi,k)) = M$
(c) $\gamma_{|L} = u_{|L}$

The construction of γ is as follows. To $x \in T$ we assign the point $y \in M$ which is closest to x. We may choose T such that this map is well defined. Let $A \in G(\Xi,k)$ be orthogonal projection onto the normal plane of M in Ξ at y. Then we set

$$\gamma(x) = (A, A(y-x)) \in E(\Xi,k).$$

Statements (a) and (b) are obvious from the construction of γ. We apply Lemma 5.3 to the variety V from the assumption of the proposition to find an equivariant entire rational function $\eta : (\Xi, V) \to (H(\Xi), G(\Xi,k))$. The entire rational function $u : \Xi \to H(\Xi) \times \Xi = \Omega$ is then defined by

$u(x) = (\eta(x), 0)$. The assumption that the germ of M at L is the germ of V at L implies that $u(x) = \gamma(x)$ for $x \in L$, so (c) holds.

In the second step we apply Lemma 5.4 to the data produced in the first step. The notation is the same, only $G(\Xi, k)$ is the variety W in 5.4. It is the conclusion of Lemma 5.4 that there are an algebraic G-variety $U \subset \Xi \times \Omega$, a smooth G-map $\varphi: (T, L) \to (\Omega, 0)$ and an equivariant entire rational function $w: U \to E(\Xi, k)$ such that

(A) $J = \{(x, \varphi(x)) \in \Xi \times \Omega \mid x \in \operatorname{Int} T\}$ is a G-invariant open subset of $\operatorname{Nonsing} U$,

(B) The G-map $\mu: T \to E(\Xi, k)$ defined by $\mu(x) = w(x, \varphi(x))$ is close to γ,

(C) $w(x, 0) = \gamma(x)$ for all $x \in L$, or in other words $\mu_{|L} = \gamma_{|L}$.

In the following we use the discussion of transversality and isotopy explained in the beginning of this section. Consider the homotopy $H'_t = (1-t)\gamma + t\mu: T \to \Omega$ between γ and μ. The image of this homotopy is near $E(\Xi, k)$. Compose H'_t with the projection of a tubular neighborhood of $E(\Xi, k)$ onto $E(\Xi, k)$. This provides an equivariant homotopy $H : T \times I \to E(\Xi, k)$ between γ and μ. This homotopy may be chosen small (because of (B)). Each $H_t : T \to E(\Xi, k)$ may be assumed to be transverse to $G(\Xi, k)$ (here we use (a)), because T is compact and transversality is an open condition.

Consider the smooth G-manifolds

$$M = \{x \in T \mid \gamma(x) \in G(\Xi, k)\} \qquad \text{(use (b))}$$

and

$$M' = \{x \in T \mid \mu(x) = w(x, \varphi(x)) \in G(\Xi, k)\}$$

The homotopy H provides a small equivariant isotopy $\overline{h}$ between M and M'. Condition (C) implies that this isotopy may be chosen such that it fixes L.

Consider

$$M \times 0 = \{(x, 0) \in T \times \Omega \mid \gamma(x) \in G(\Xi, k))\}$$

and

$$\begin{aligned} Z' &\overset{\text{Def}}{=} J \cap w^{-1}(G(\Xi, k)) \\ &= \{(x, \varphi(x)) \in J \mid w(x, \varphi(x)) \in G(\Xi, k)\} \end{aligned}$$

The isotopy $\overline{h}$ induces a small isotopy between $M \times 0$ and $M' \times 0$ which fixes L. Because φ may be chosen small (see 5.4 (6)) there is a small isotopy between $M' \times 0$ and Z'. The isotopy fixes L because φ vanishes on L. The composition of these two small isotopies provides a small isotopy h_t between $M \times 0$ and Z'. The isotopy extends to a small ambient isotopy h_t of $\Xi \times \Omega$ in the C^1 topology. In particular, $M \times 0 = h_0(M \times 0)$ and $Z' = h_1(M \times 0)$.

We show that Z' is the union of non-singular components of a G-variety Z. We summarize the required information. Let

$$Z'' = U \cap w^{-1}(G(\Xi, k)).$$

We constructed an equivariant entire rational function $w: U \to E(\Xi, k)$ between these algebraic G-varieties. Let $z \in Z' = J \cap w^{-1}(G(\Xi, k))$. Then

(1) $w(z)$ is non-singular of dimension $\dim G(\Xi, k)$ in $G(\Xi, k)$
(2) z is non-singular of dimension n in U (see 5.4 (5) and use (A))
(3) z is non-singular of dimension $k + \dim G(\Xi, k)$ in $E(\Xi, k)$
(4) w is transverse to $G(\Xi, k)$ at z.

These four statements imply (see [**AK1**, Lemma 1.4]) that points of Z' are non-singular of dimension $n - k = \dim M$ in Z''. Taking Sing a number of times we construct the variety (see 4.4 and use 4.5)

$$Z = \operatorname{Sing}(\operatorname{Sing}(\ldots(\operatorname{Sing} Z'')\ldots))$$

so that $\dim Z = \dim M$, and Z' is the union of components of Z. Since we have a small G-isotopy h_t of $\Xi \times \Omega$ which sends $M \times 0$ to Z' and fixes $L \times 0$, the proposition is proved. ∎

After these preparations we shall complete the proof of Theorem 1.3. The argument is almost the same as the second part of the proof of [**AK1**, Proposition 2.8].

Proof of Theorem 1.3: By assumption there is a G-cobordism N between M and a non-singular algebraic G-variety L. We may embed N equivariantly in a representation Ξ of G, such that L is realized as zero set of a G-invariant polynomial. Let $Y = \partial(N \times [-1, 1]) \subset \Xi \times \mathbf{R}$. (We round off the corners of Y without changing the embedding near $N \times 0$. Without changing notation, this slight modification of Y will be understood in the following.) The germ of Y at $L \times 0$ is the germ of the non-singular algebraic G-variety $L \times \mathbf{R}$ at $L \times 0$. Proposition 5.5 applied to the closed G-submanifold Y of $\Xi \times \mathbf{R}$ and the non-singular G-variety $L \subset Y$ provides us with a representation Ω of G, a G-isotopy h_t of $\Xi \times \mathbf{R} \times \Omega$, and an algebraic G-variety Z such that $Z' = h_1(Y \times 0)$ consists of a union of non-singular components of Z.

Decompose Y as $Y_+ \cup Y_-$ where

$$Y_\pm = \{(x,t) \in \Xi \times \mathbf{R} \mid (x,t) \in Y \text{ and } \pm t \geq 0\}$$

Set $Y_0 = Y_+ \cap Y_- = M \times 0 \cup L \times 0$. Accordingly, Z' decomposes into $Z'_\pm = h_1(Y_\pm \times 0)$. Their intersection $h_1(Y_0 \times 0)$ separates Z compactly and equivariantly setting $Z_0 = Z'_- \cup (Z - Z')$ and $Z_1 = Z'_+$. By construction, Y_0 is a codimension 1 G-submanifold of Z. By Lemma 5.2 there are small G-isotopies H_t of Nonsing Z which fix $L \times 0 \times 0$ and carry $h_1(Y_0 \times 0)$ to a non-singular algebraic G-variety. So $H_1(h_1(Y_0 \times 0))$ is a non-singular algebraic G-variety and $L \times 0 \times 0$ is a subvariety. Let

$$V = H_1(h_1(M \times 0 \times 0)) = H_1(h_1(Y_0 \times 0)) - (L \times 0 \times 0)$$

The algebraic G-varieties $H_1(h_1(Y_0 \times 0))$ and $L \times 0 \times 0$ have the same dimension and $L \times 0 \times 0$ is a closed subvariety in $H_1(h_1(Y_0 \times 0))$. It follows from these properties (see Lemma 1.6 of [**AK1**]) that their difference V is a non-singular algebraic G-variety. By construction it is isotopic to $M \times 0 \times 0$ through a small G-isotopy. This proves Theorem 1.3. ∎

6. More Groups which Act Without Fixed Point

Let G be a compact Lie group, H a subgroup of finite index k, and let V be an H space. We denote the set of H maps from G to V by $\mathrm{Map}^H(G,V)$. This set has a G-action defined by $(g\varphi)(g') = \varphi(gg')$. As a space, $W = \mathrm{Map}^H(G,V)$ is isomorphic to the k-fold cartesian product of V.

Proposition 6.1.

(1) *If* $V^H = \emptyset$ *then* $W^G = \emptyset$
(2) *If* V *is diffeomorphic to* $\mathbf{R}^n$, *then* W *is diffeomorphic to* $\mathbf{R}^{nk}$.
(3) *If* V *is an algebraic* H*-variety, then* W *is an algebraic* G*-variety.*

PROOF: (1) follows because $W^H = V^H \times \cdots \times V^H$, and (2) is clear. Let Γ be a representation of H in which V is realized as zero set. Then $V \times \cdots \times V = \mathrm{Map}^H(G,V) \subset \mathrm{Map}^H(G,\Gamma) = \Gamma \times \cdots \times \Gamma$. Clearly this is an inclusion as a subvariety into the representation $\mathrm{Map}^H(G,\Gamma)$ of G. ∎

Proposition 6.2. *Suppose that there is an epimorphism* $G \to K$ *and that there is a* K*-variety* X *diffeomorphic to* $\mathbf{R}^n$ *such that* $X^K = \emptyset$. *Then there is a* G*-variety* Y *diffeomorphic to* $\mathbf{R}^m$ *such that* $Y^G = \emptyset$. *(All actions are assumed to be effective.)*

PROOF: The map from G to K induces an action of G on X. It is algebraic. Let Z be a faithful representation of G and $Y = X \times Z$ with diagonal action. This space has all of the desired properties. ∎

References

[**AK1**] S. Akbulut and H. King, *The Topology of Real Algebraic Sets with Isolated Singularities*, Annals of Mathematics **113** (1981), 425–446.

[**AK2**] S. Akbulut and H. King, *A Relative Nash Theorem*, Transactions of the Amer. Math. Soc. **267** (1981), 465–481.

[**AS**] M. F. Atiyah and I. M. Singer, *The Index of Elliptic Operators, III*, Annals of Mathematics **87** (1968), 546–604.

[**Ba**] A. Bak, *The Computation of Surgery Groups of Finite Groups with Abelian 2-Hyperelementary Subgroups*, in Algebraic K-Theory, Evanston 1976, Lecture Notes in Mathematics **551** (1976), 384–409.

[**BH**] H. Bass and W. Haboush, *Linearizing Certain Reductive Group Actions*, Transactions of the Amer. Math. Soc. **292** (1984), 463–482.

[**BCR**] J. Bochnak, M. Coste and M-F. Roy, "Géométrie Algébrique Réelle," Ergebnisse der Mathematik und ihrer Grenzgebiete, Springer-Verlag, 1987.

[**B**] G. Bredon, "Introduction to Compact Transformation Groups," Academic Press, 1972.

[**BKS**] N. P. Buchdahl, S. Kwasik, and R. Schultz, *One Fixed Point Actions on Low-Dimensional Spheres*, preprint 1989.

[**CF**] P. E. Conner and E. E. Floyd, *On the Construction of Periodic Maps without Fixed Points*, Proc. Amer. Math. Soc. **10** (1959), 354–360.

[**CM**] P. E. Conner and D. Montgomery, *An Example for* SO(3), Proc. Nat. Acad. Sci. U. S. A. **48** (1962), 1918–1922.

[**D**] K. H. Dovermann, *Almost Isovariant Normal Maps*, to appear in Amer. J. of Math..

[**DR1**] K. H. Dovermann and M. Rothenberg, *Equivariant Surgery and Classification of Finite Group Actions on Manifolds*, Memoirs of the Amer. Math. Soc. **397** (1988).

[**DR2**] K. H. Dovermann and M. Rothenberg, *The Generalized Whitehead Torsion of a G-Fibre Homotopy Equivalence*, to appear in the proceedings of the International Conference on Transformation Groups in Osaka (1987).

[**Dr**] A. Dress, *Induction and Structure Theorems for Orthogonal Representations of Finite Groups*, Annals of Mathematics (1975), 291–325.

[**EL**] A. L. Edmonds and R. Lee, *Compact Lie Groups which Act on Euclidean Space without Fixed Points*, Proc. Amer. Math. Soc. **55** (1976), 416–418.

[**F**] M. Furuta, *A Remark on Fixed Points of Finite Group Actions on* S^4, Topology **28** (1989), 35–38.

[**H**] M. W. Hirsch, "Differential Topology," Graduate Texts in Mathematics Vol. 33, Springer-Verlag, New York, Heidelberg, Berlin, 1976.

[**HH**] W. C. Hsiang and W. Y. Hsiang, *Differentiable Group Actions of Connected Classical Groups, I*, Amer. J. of Math. **89** (1967), 705–786.
[**I**] N. V. Ivanov, *Approximation of Smooth Manifolds by Real Algebraic Sets*, Russian Math. Surveys **37:1** (1982), 1–59.
[**Ka**] T. Kambayashi, *Automorphism Group of a Polynomial Ring and Algebraic Group Actions on an Affine Space*, Jour. of Algebra **60** (1979), 439–451.
[**K**] H. King, *Approximating Submanifolds of Real Projective Space by Varieties*, Topology **15** (1976), 81–85.
[**Kr**] H. Kraft, *Algebraic Automorphisms of Affine Space*, In this volume.
[**M1**] J. Milnor, "Singular Points of Complex Hypersurfaces," Annals of Mathematics Studies, Princeton University Press, Princeton N.J., 1968.
[**M2**] J. Milnor, *On the Stiefel-Whitney Numbers of Complex Manifolds and of spin-Manifolds*, Topology **3** (1965), 223–230.
[**MS**] D. Montgomery and H. Samelson, *Fiberings with Singularities*, Duke Math. J. **13** (1946), 51–56.
[**Mo1**] M. Morimoto, *On one Fixed Point Actions on Spheres*, Proc. of the Japan Academy **63** (1987), 95–97.
[**Mo2**] M. Morimoto, S^4 *does not have one Fixed Point Actions*, Osaka J. of Math. **25** (1988), 575–580.
[**N**] J. Nash, *Real Algebraic Manifolds*, Annals of Math. **56** (1952), 405–421.
[**O**] R. Oliver, "Whitehead Groups of Finite Groups," London Mathematical Society Lecture Note Series 132, Cambridge University Press, 1988.
[**P1**] T. Petrie, *One Fixed Point Actions on Spheres, I*, Advances in Mathematics **46** (1982), 3–14.
[**P2**] T. Petrie, *One Fixed Point Actions on Spheres, II*, Advances in Mathematics **46** (1982), 15–70.
[**PR1**] T. Petrie and J. Randall, *Finite-order Algebraic Automorphisms of Affine Varieties*, Comment. Math. Helvetici **61** (1986), 203–221.
[**PR2**] T. Petrie and J. Randall, "Transformation Groups on Manifolds," Dekker Lecture Series 82, Marcel Dekker, 1984.
[**RU**] I. Reiner and S. Ullom, *Remarks on the Class Groups of Integral Group Rings*, Symposia Mathematica (Academic Press) **XIII** (1974), 501–516.
[**S1**] G. Schwarz, *Smooth Functions Invariant under the Action of a Compact Lie Group*, Topology **14** (1975), 63–68.
[**S2**] G. Schwarz, *Exotic Algebraic Group Actions*, C. R. Acad. Sci. (1989) (to appear).
[**Se**] H. Seifert, *Algebraische Approximation von Mannigfaltigkeiten*, Math. Zeitschrift **41** (1936), 1–17.
[**Sh**] J. Shaneson, *Wall's Surgery Obstruction Groups for* $\mathbf{Z} \times G$, Ann. of Math. **90** (1969), 296–334.
[**St**] E. Stein, *Surgery on Products with Finite Fundamental Group*, Topology **16** (1977), 473–493.
[**T**] A. Tognoli, *Su una Congettura di Nash*, Annali della Scuola Normale Superiore di Pisa **27** (1973), 167–185.

[**W1**] C. T. C. Wall, "Surgery on Compact Manifolds," Academic Press, 1970.
[**W2**] C. T. C. Wall, *Classification of Hermitian Forms. VI Group Rings*, Annals of Mathematics **103** (1976), 1–80.

Karl Heinz Dovermann
Department of Mathematics
University of Hawaii
Honolulu, HI 96822
USA

Mikiya Masuda
Department of Mathematics
Osaka City University
Osaka
JAPAN

Ted Petrie
Department of Mathematics
Rutgers University
New Brunswick, NJ 08903
USA

ALGEBRAIC AUTOMORPHISMS OF AFFINE SPACE

HANSPETER KRAFT*

Table of Contents

Introduction

The following article is an updated version of the report [**Kr85**]; for convenience of the reader we have incorporated most of the material of that paper. Some other aspects can be found in the article [**Ba85**] of BASS.

0.1. Our main concern is the following fundamental problem in the theory of algebraic transformation groups:

> *How can a complex algebraic group G act algebraically on affine space $\mathbf{C}^n$?*

Of course, the most obvious actions are the *linear* ones, given by *rational representations* $\rho : G \rightarrow \mathrm{GL}(V)$ where "rational" means that ρ is algebraic. Let us call an action *linearizable* if it is obtained from a rational

*Partially supported by SNF (Schweizerischer Nationalfonds)

representation by a polynomial change of coordinates. This means that there is a G-equivariant algebraic isomorphism $\mathbf{C}^n \xrightarrow{\sim} V$ where V is a G-module.

The example of the additive group $\mathbf{C}^+$ acting by translations $x \mapsto x + tv$ ($x \in \mathbf{C}^n, t \in \mathbf{C}^+$) where v is a fixed vector, shows that an action of a *unipotent* group need not be linearizable. At most one could expect that such actions are *triangularizable*; but this turns out to be wrong in general (see 1.4).

0.2. For *reductive* groups the situation is more difficult. In the last few years a lot of work has been invested in the following problem (see § 5; a positive response has been conjectured by KAMBAYASHI [**Ka79**, Conjecture 3.1]):

Linearization Problem. *Is any action of a reductive complex algebraic group G on affine space linearizable?*

Very recently SCHWARZ gave the first examples of non-linearizable actions on affine space [**Sch89**]. Among other things he showed that there are *infinitely many non-equivalent actions of* O_2 *on* $\mathbf{C}^4$, *of* SL_2 *on* $\mathbf{C}^7$, and of SO_3 on $\mathbf{C}^{10}$. These examples came as a real surprise; they disproved several conjectures made in this context (see § 6).

0.3. If an action on $\mathbf{C}^n$ is linearizable the fixed point set $(\mathbf{C}^n)^G$ is isomorphic to some $\mathbf{C}^d$ and in particular non-empty. So we may ask the following question:

Fixed Point Problem. *Given an action of a reductive complex algebraic group G on $\mathbf{C}^n$, is the fixed point set $(\mathbf{C}^n)^G$ isomorphic to $\mathbf{C}^d$, or at least non-empty?*

Several well-known results from smooth transformation groups carry over to algebraic actions (see § 3). But at the moment we don't know of any example of a reductive group action on affine space without fixed points.

In the following we will describe some results in this context and indicate a number of related problems.

Convention. Our base field is the field $\mathbf{C}$ of complex numbers; we always work in the category of algebraic varieties over $\mathbf{C}$. Of course, we could replace $\mathbf{C}$ by any other algebraically closed field of characteristic zero. Some results even hold in positive characteristic, and there are also interesting *rationality* questions (cf. [**Sh**] in this volume). But we have not attempted to work out the most general setting.

§ 1 Affine Cremona Group

1.1. We denote by $\mathcal{G}_n$ the group of *algebraic automorphisms* of the affine space $\mathbf{C}^n$. Such an automorphism $\varphi : \mathbf{C}^n \xrightarrow{\sim} \mathbf{C}^n$ is given by an n-tuple $(\varphi_1, \ldots, \varphi_n)$ of polynomials $\varphi_i \in \mathbf{C}[X_1, \ldots, X_n]$ generating the polynomial ring: $\mathbf{C}[\varphi_1, \ldots, \varphi_n] = \mathbf{C}[X_1, \ldots, X_n]$. The *degree* of φ is defined by $\deg \varphi := \max_i \deg \varphi_i$. It is well known that a polynomial map $\varphi : \mathbf{C}^n \to \mathbf{C}^n$ is an isomorphism if and only if φ is bijective:

$$\mathcal{G}_n = \{\varphi : \mathbf{C}^n \xrightarrow{\text{bij}} \mathbf{C}^n \mid \varphi \text{ polynomial}\} \simeq \operatorname{Aut} \mathbf{C}[X_1, \ldots, X_n].$$

$\mathcal{G}_n$ is a subgroup of the automorphism group of the field of *rational functions* $\mathbf{C}(X_1, \ldots, X_n)$ which is classically called the *Cremona group* and corresponds geometrically to the *rational* automorphisms of $\mathbf{C}^n$. We will use the name *affine Cremona group* for $\mathcal{G}_n$.

$\mathcal{G}_n$ contains two important subgroups: the subgroup $\mathcal{A}_n$ of *affine transformations* and the *Jonquière subgroup* $\mathcal{J}_n$ of *triangular transformations*, defined in the following ways:

$$\begin{aligned}\mathcal{A}_n &:= \{\varphi = (\varphi_1, \ldots, \varphi_n) \in \mathcal{G}_n \mid \varphi_i \text{ linear } \forall i\},\\ \mathcal{J}_n &:= \{\varphi = (\varphi_1, \ldots, \varphi_n) \in \mathcal{G}_n \mid \varphi_i \in \mathbf{C}[X_1, \ldots, X_i] \;\forall i\}.\end{aligned}$$

Clearly $\mathcal{A}_n$ is the semidirect product of the general linear group GL_n with the subgroup $\mathcal{T}_n$ of *translations*. Moreover, any element $\varphi = (\varphi_1, \ldots, \varphi_n)$ of the Jonquière subgroup $\mathcal{J}_n$ has the form

$$\varphi_i(X_1, \ldots, X_n) = \lambda_i X_i + \psi_i(X_1, \ldots, X_{i-1})$$

where $\lambda_i \in \mathbf{C}^*$. This is an easy consequence of the fact that the *Jacobian* $\operatorname{Jac} \varphi := \det(\frac{\partial \varphi_i}{\partial x_j})$ of any algebraic isomorphism φ is a constant $\neq 0$.

At this point we remark that the famous *Jacobian conjecture* is still completely open, even in the two dimensional case (cf. [**BCW82**]). It states that any polynomial map $\varphi : \mathbf{C}^n \to \mathbf{C}^n$ with $\operatorname{Jac} \varphi = const \neq 0$ is an isomorphism.

1.2. Shafarevich shows in [**Sh66**] that the group $\mathcal{G}_n$ has the structure of an ∞-*dimensional algebraic group*. This means that $\mathcal{G}_n$ can be written as a union $\mathcal{G}_n = \bigcup_{j=1}^{\infty} Y_j$ of algebraic varieties Y_j with the property that Y_j is a closed subvariety of Y_{j+1} for all j and that multiplication and taking inverse are algebraic morphisms with respect to the Y_j's (cf. [**Ka79**]). For example we may define Y_j to be the set of all automorphisms φ of degree $\leq j$. In particular, one has the notion of *algebraic subgroups* of $\mathcal{G}_n$, i.e.

those subgroups which are contained in one of the Y_j's as a subvariety. It is not difficult to show that the Linearization Problem 0.2 is equivalent to the following:

Conjugation Problem. *Is any reductive algebraic subgroup of the affine Cremona group $\mathcal{G}_n$ conjugate to a subgroup of* GL_n*?*

As mentioned above the counterexamples of SCHWARZ in [**Sch89**] are also counterexamples to the Conjugation Problem; e.g. there are subgroups of $\mathcal{G}_n$ isomorphic to SL_2 which are not conjugate to a subgroup of GL_n.

1.3. Not much is known about the structure of $\mathcal{G}_n$ in general except for $n = 2$ and of course for the trivial case $n = 1$ where $\mathcal{G}_1 = \mathcal{A}_1$. We omit this case in the sequel. The following important result goes back to VAN DER KULK [**Ku53**] (cf. [**Na72**]). A modern proof based on SERRE's theory of *trees* can be found in [**GD75**].

Theorem. *$\mathcal{G}_2$ is the amalgamated product $\mathcal{A}_2 *_{\mathcal{B}_2} \mathcal{J}_2$ where $\mathcal{B}_2 := \mathcal{A}_2 \cap \mathcal{J}_2$.*

This strong structure theorem has several important consequences. The following one covers and generalizes a number of special results about algebraic group actions on $\mathbf{C}^2$ obtained with different methods by GUTWIRTH, BIALYNICKI-BIRULA, RENTSCHLER, MIYANISHI, IGARASHI, FURUSHIMA (see [**Gu62**], [**BB67**], [**Re68**], [**Mi71**], [**Ig77**], [**Fr82**]).

Corollary 1. *Every action of an algebraic group on $\mathbf{C}^2$ is conjugate to either an affine action or a triangular action.*

In fact, one shows that any algebraic subgroup of $\mathcal{G}_2 = \mathcal{A}_2 *_{\mathcal{B}_2} \mathcal{J}_2$ is of *bounded length.* (Recall that the length of an element $g \in \mathcal{A}_2 *_{\mathcal{B}_2} \mathcal{J}_2$ is the minimal number of elements in $\mathcal{A}_2 \cup \mathcal{J}_2$ needed to express g.) Now we apply a general result about such subgroups of amalgamated products stating that they are conjugate to subgroups of one of the factors (see [**Se80**, I.4.3, Theorem 8]). ■

As a consequence we obtain a positive answer to the Linearization Problem in dimension 2 (cf. [**Ka79**]):

Corollary 2. *Every reductive group action on $\mathbf{C}^2$ is linearizable.*

1.4. In general one does not know if $\mathcal{G}_n$ is generated by $\mathcal{A}_n$ and $\mathcal{J}_n$. However, an amalgamated product structure for $\mathcal{G}_n$ as in the theorem above *does not exist for $n \geq 3$*. In fact, such a structure theorem for $\mathcal{G}_n$ would imply the analogue of Corollary 1 for algebraic group actions on $\mathbf{C}^n$. But BASS [**Ba84**] has constructed a $\mathbf{C}^+$-action on $\mathbf{C}^3$ which is not *triangularizable*, i.e. which cannot be factored through the Jonquière subgroup or through the affine subgroup: He considers the vector field

$$D := (xz + y^2)(x\frac{\partial}{\partial y} - 2y\frac{\partial}{\partial z})$$

on $\mathbf{C}^3$. It is easy to see that D is locally nilpotent, hence determines a $\mathbf{C}^+$-action on $\mathbf{C}^3$ (via exponentiation). The fixed point set of this action coincides with the zero set of the vector field D: It is the quadratic surface $S : xz + y^2 = 0$ which has an isolated singularity at 0. This shows that the action is not triangularizable.

The construction of BASS has been generalized by POPOV [**Po87**] (see the report of SNOW [**Sn**] in this volume).

§ 2 Cancellation and Characterization of $\mathbf{C}^n$

For a more detailed discussion of the following we refer to the article [**Su**] of SUGIE in this volume.

2.1. We now state another problem sometimes called Zariski's problem, which is strongly related to the questions above.

Cancellation Problem. *Given a variety Y and an isomorphism $Y \times \mathbf{C}^m \xrightarrow{\sim} \mathbf{C}^n$, for some $m, n \in \mathbf{N}$, does this imply that Y is isomorphic to $\mathbf{C}^{n-m}$?*

It is easy to see that such a variety Y is *affine, smooth* and *unirational*, and that every *invertible* algebraic function on Y is a *constant*. In case $\dim Y = 1$ this already implies that $Y \simeq \mathbf{C}$. Furthermore, any such variety Y is *factorial*; more generally, every *vector bundle* on Y is *trivial*. An obvious topological property is that Y is *contractible* (with respect to the $\mathbf{C}$-topology).

2.2. Besides the easy case of dimension 1 just mentioned, one also has a positive answer for the Cancellation Problem for $\dim Y = 2$, due to the work of FUJITA, MIYANISHI and SUGIE (cf. [**Fu79**]). They gave the following characterization of the affine plane (see [**MS80**], [**Ka80**], [**Ru81**]):

Theorem. *Let Y be a smooth factorial affine surface. If there is a dominant morphism $\varphi : \mathbf{C}^n \to Y$ for some $n \in \mathbf{N}$, then Y is isomorphic to $\mathbf{C}^2$. In particular, the Cancellation Problem has a positive answer in dimension 2.*

This result has interesting applications to the structure of algebraic quotients of reductive group actions (see Proposition 4.2).

2.3. Remarks. (a) There is a *topological* characterization of $\mathbf{C}^2$ which was obtained by RAMANUJAM [**Ra71**]:

> *A non-singular affine surface which is contractible and simply connected at infinity is isomorphic to* $\mathbf{C}^2$.

He also gave an example of a contractible non-singular surface which is *not isomorphic* to $\mathbf{C}^2$. In the last few years the study of *acyclic* and *contractible* surfaces has made considerable progress due to the work of MIYANISHI and TOM DIECK-PETRIE (cf. [**DP89**] in this volume).

(b) It is an open question whether the analogue of Theorem 2.2 holds in arbitrary dimension. There is an "algebro-topological" characterization of $\mathbf{C}^3$ due to MIYANISHI [**Mi84**].

(c) Work of BEAUVILLE, COLLIOT-THÉLÈNE, SANSUC, SWINNERTON-DYER [**BC85**] has shown the *birational cancellation* to be false:

> *They construct a 3-dimensional non-rational variety* Y *such that* $Y \times \mathbf{P}^3$ *is rational.*

2.4. The Cancellation Problem is strongly related with the problems discussed so far. It is easy to see that a positive solution of the Linearization Problem 0.2 (or even of the Fixed Point Problem 0.3 in its strong form) would have implied a positive answer for the Cancellation Problem: Consider the $\mathbf{Z}/2$ - action on $Y \times \mathbf{C}^m \simeq \mathbf{C}^n$ given by $(x, y) \mapsto (x, -y)$. Then Y is the fixed point set, hence isomorphic to some $\mathbf{C}^d$. (Of course we could argue in the same way using a torus or any other reductive group instead of $\mathbf{Z}/2$, choosing a representation on $\mathbf{C}^m$ or on some $\mathbf{C}^{m'}$ with $m' \geq m$ which does not contain the trivial representation.)

2.5. Generalization. The Cancellation Problem has the following generalization:

> *Given two affine varieties* Y *and* Z *and an isomorphism* $Y \times \mathbf{C}^k \simeq Z \times \mathbf{C}^k$, *does this imply that* Y *and* Z *are isomorphic?*

Recently DANIELEWKI showed that the answer is no. In fact, there is the following result (see [**Da89**]):

Proposition. *For* $n \geq 1$ *let* $Y_n \subset \mathbf{C}^3$ *be the closed subvariety defined by* $x^n y + z^2 = 1$.

(a) *The varieties* $Y_n \times \mathbf{C}$ *are all isomorphic.*

(b) (FIESELER) *The topological spaces Y_n are all of different homeomorphism type.*

In the construction of DANIELEWKI the varieties Y_n appear as principal $\mathbf{C}^+$-bundles over the prevariety $\tilde{\mathbf{C}} := \mathbf{C} \cup_{\dot{\mathbf{C}}} \mathbf{C}$—the line with a point doubled, obtained by glueing two copies of $\mathbf{C}$ along $\dot{\mathbf{C}} := \mathbf{C} \setminus \{0\}$. From this one easily deduces the first assertion. Concerning the second, FIESELER showed that the first homology group at infinity of Y_n is cyclic of order $2n$.

§ 3 Fixed Point Sets

3.1. We now discuss some results concerning the Fixed Point Problem 0.3. We first remark that the reductivity assumption is essential:

For any non-reductive group G there is an action on some $\mathbf{C}^n$ without fixed points.

To see this we choose a representation V containing a subrepresentation W of codimension 1 which has no G-stable complement; such a pair is easily constructed from any representation which is not completely reducible (see [**KP85**, Note added in proof]). Then the action of G on the affine space $\mathbf{P}(V) \setminus \mathbf{P}(W)$ has the required properties.

Next we remark that the fixed point set $F = (\mathbf{C}^n)^G$ of a reductive group G is always a *smooth* closed subvariety. This holds more generally for an action of G on any *smooth* variety (see [**Fo73**], or [**Lu73**]), but is not true for non-reductive groups as shown by the example of BASS given in 1.4.

3.2. A reductive group G contains a *compact* subgroup $K \subset G$ (with respect to the $\mathbf{C}$-topology) which is *Zariski-dense*. K is uniquely determined up to conjugation. It follows that $X^G = X^K$ for any G-variety X, i.e. any variety X with an algebraic action of G. In particular, we can apply methods from the theory of compact transformation groups (cf. BREDON's book [**Br72**]). However, the condition of an action to be algebraic is very restrictive and has important consequences which do not hold in the topological setting. For example, for any algebraic action of a finite group G on a variety X the fixed point set of any subgroup $H \subset G$ is a subvariety, hence has the homotopy type of a finite CW-complex (cf. [**KP85**, Appendix]).

3.3. The following results of PETRIE and RANDALL [**PR86**] give some more properties of algebraic actions which do not hold in the differentiable setting.

Theorem 1. *If the finite group G acts algebraically on a smooth affine variety X, then X has the equivariant homotopy type of a finite G-CW-complex.*

From this they obtain a *Lefschetz fixed point formula* which leads to the following consequence (cf. VERDIER [**Ve73**]):

Corollary. *Assume that the finite group G has a normal series $P \subset H \subset G$ where P and G/H are both groups of prime power order and H/P is cyclic. If G acts algebraically on an acyclic variety X, then X contains a fixed point.*

Example. An algebraic action of the *binary octahedral group* O on $\mathbf{C}^n$ always has a fixed point. In particular, we get the following result due to PANYUSHEV [**Pa84**, 3.2 Theorem 7]:

> *If* $\mathrm{SL}_2(\mathbf{C})$ *acts on* $\mathbf{C}^n$ *without three-dimensional orbits, then it has a fixed point.*

(Use the fact that O is a maximal subgroup of $\mathrm{SL}_2(\mathbf{C})$.)

Theorem 2. *If a finite group G acts on a non-singular affine variety X, then X admits a G-compactification.*

(Here a G-compactification means a compact smooth G-manifold X^* with boundary such that $X \subset X^*$ and $\partial X^* = X^* \setminus X$.)

3.4. The famous *Smith-theory* provides us with other important tools. They can be used for finite p-groups but hold in a much more general setting (cf. [**Br72**]). A typical result is the following one:

Proposition 1. *Let G be a finite p-group and X a G-variety. Assume that $H^*(X, \mathbf{Z}/p) = \mathbf{Z}/p$. Then $H^*(X^G, \mathbf{Z}/p) = \mathbf{Z}/p$. In particular, the fixed point set X^G is non-empty and connected.*

We can apply this to torus actions on $\mathbf{C}^n$ using the following two facts:

(a) For any prime p the elements of p-power order in T form a Zariski-dense subgroup $T_p \subset T$; hence $(\mathbf{C}^n)^{T_p} = (\mathbf{C}^n)^T$.

(b) There is a finite subgroup $T' \subset T_p$ with the same fixed points as T_p. (Remember that $\mathbf{C}^n$ is quasi-compact in the Zariski-topology.)

Proposition 2. *Let T be a torus acting algebraically on $\mathbf{C}^n$. Then we have $H^*((\mathbf{C}^n)^T, \mathbf{Z}) = \mathbf{Z}$. In particular, the fixed point set is non-empty and connected.*

(Cf. BIALYNICKI-BIRULA [**BB67**]; the result also follows directly from [**Br72**, III.10] using the Zariski-dense compact torus in T.)

Remark. As a consequence we see that a 1-dimensional fixed point set of a p-group or a torus acting on $\mathbf{C}^n$ is always isomorphic to the affine line $\mathbf{C}$. Is this line ambient isomorphic (i.e. by an automorphism of $\mathbf{C}^n$) to a coordinate line? For $n = 2$ this is the famous result of ABHYANKAR-MOH [**AM75**], and for $n \geq 4$ it follows from [**Je87**, Theorem 1.1]. What can be said about a two-dimensional fixed point set? Is it isomorphic to the plane $\mathbf{C}^2$?

3.5. The next result is obtained with methods similar to those used by OLIVER in his paper [**Ol79**] about differentiable $\mathrm{SO}_3(\mathbf{R})$-actions on disks.

Proposition. *Any* $\mathrm{SL}_2(\mathbf{C})$*-action on* $\mathbf{C}^n$ *has a fixed point for* $n \leq 7$.

OLIVER also gives an example of an $\mathrm{SO}_3(\mathbf{R})$-action on the 8-dimensional disk without fixed points!

3.6. The Fixed Point Problem for *smooth actions of compact semi-simple groups* was studied systematically by HSIANG and STRAUME (see [**HS82**, **HS86**]). It should be an interesting task to generalize their results to the algebraic setting. One step in this direction (besides Proposition 3.5 above) is Theorem 7.3 which has been obtained in collaboration with LUNA. It states that any action of a reductive group G on $\mathbf{C}^n$ has fixed points if there are not too many invariant functions, i.e. if the quotient $\mathbf{C}^n /\!\!/ G$ (4.1) has dimension ≤ 1.

§ 4 Algebraic Quotients

Before continuing our report we want to indicate some general methods from algebraic transformation groups and invariant theory which are useful for the study of group actions on affine spaces and which will be needed in the sequel.

4.1. Invariant rings and quotients. Let G be a reductive group acting on an affine variety X. Denote by $\mathcal{O}(X)$ the algebra of regular functions on X. The general *finiteness theorem* of HILBERT states that the algebra $\mathcal{O}(X)^G$ of G-*invariant functions* on X is *finitely generated* (see [**Kr84**,

II.3.2]). Hence we can define an affine variety $X/\!\!/G := \operatorname{Spec} \mathcal{O}(X)^G$, called the *algebraic quotient* (or categorical quotient) of X by G, and a map $\pi : X \to X/\!\!/G$, the *quotient map*, induced by the inclusion $\mathcal{O}(X)^G \subset \mathcal{O}(X)$. For a proof of the following results see [**Kr84**, II.3.2] or [**MF82**, Chap. 1, § 2]:

Proposition. *The quotient map $\pi : X \to X/\!\!/G$ has the following properties:*

(a) *π is universal with respect to all morphisms $\varphi : X \to Y$ which are constant on orbits;*
(b) *π is surjective and submersive (i.e. $X/\!\!/G$ carries the quotient topology);*
(c) *π is G-closed, i.e. the image of a G-stable closed subvariety is closed;*
(d) *π is G-separating, i.e. it separates disjoint G-stable closed subvarieties.*

In particular, we see that each fibre $\pi^{-1}(y)$, $y \in X/\!\!/G$, contains exactly one *closed* orbit O_y and that

$$\pi^{-1}(y) = \{x \in X \mid \overline{Gx} \supset O_y\}.$$

(Remember that $\overline{Gx} \setminus Gx$ is closed in $\overline{Gx}$ and of strictly smaller dimension.) This shows that the quotient parametrizes the *closed orbits* in X. Hence, if all orbits are closed (e.g. if G is finite), then the quotient is the orbit space X/G and is called the *geometric quotient.* In general it is the "best algebraic approximation" to the orbit space X/G.

Remark. All varieties defined over the complex numbers $\mathbf{C}$ carry another topology besides the Zariski-topology, which is induced by the usual topology of $\mathbf{C}^n$. We call it the $\mathbf{C}$-*topology.* It was shown by LUNA [**Lu75b**] that all the properties above also hold with respect to this stronger topology (4.3; cf. [**Sch**]).

4.2. Structure of quotients. In general, it is rather difficult to determine the structure of the quotient $X/\!\!/G$, even if we start with a representation $X = V$ of G. Some properties are inherited from X, like *normality* and also *factoriality* in case G has a trivial character group (e.g. if G is semisimple and connected). Furthermore, the theorem of BOUTOT states that X has *rational singularities* if X does [**Bt87**]. In connection with the *Picard group*, we have the following lemma (cf. [**DMV89**]):

Lemma. *The canonical map $\pi^* : \operatorname{Pic} X/\!\!/G \to \operatorname{Pic} X$ is injective in the following cases:*

(a) *G is connected and every maximal torus has a fixed point;*

(b) *G has a fixed point and every invertible function on X is constant.*

Applying all this to actions on affine spaces we get a number of important results about the structure of the quotient, which we collect in the following proposition.

Proposition. *Let G be a reductive group acting on $\mathbf{C}^n$.*

(a) *$\mathbf{C}^n/\!\!/G$ has rational singularities. In particular, it is normal and Cohen-Macaulay.*

(b) $\operatorname{Pic} \mathbf{C}^n/\!\!/G = 1$; *hence $\mathbf{C}^n/\!\!/G$ is factorial if and only if it is locally factorial.*

(c) *If $\mathbf{C}^n/\!\!/G$ is smooth and of dimension 2, then $\mathbf{C}^n/\!\!/G \simeq \mathbf{C}^2$.*

(d) *If G is semisimple and connected and $\dim \mathbf{C}^n/\!\!/G = 2$, then $\mathbf{C}^n/\!\!/G \simeq \mathbf{C}^2$.*

The assertion (c) follows from the characterization of $\mathbf{C}^2$ given in 2.2; for (d) one uses in addition a result of KEMPF's about the smoothness of 2-dimensional quotients [**Ke80**].

4.3. Topology of quotients. [1] For a compact Lie group K acting on a topological space T quite a lot is known about the orbit space T/K (cf. BREDON's book [**Br72**]). In the work [**KPR86**] of KRAFT, PETRIE AND RANDALL some of these results have been extended from the topological to the algebraic setting by comparing the orbit space X/K with the quotient $X/\!\!/G$; here G is a reductive algebraic group acting algebraically on an affine variety X and $K \subset G$ is a maximal compact subgroup. The following are two of the main results:

Theorem 1. *The canonical map $\bar{\pi} : X/K \to X/\!\!/G$ induces an isomorphism in cohomology.*

Theorem 2. *If X is contractible (respectively acyclic), then so is $X/\!\!/G$.*

(Theorem 2 follows from Theorem 1 by using OLIVER's work about the *Conner Conjecture* [**Ol76**].) In particular, we see that any quotient of the form $\mathbf{C}^n/\!\!/G$ is *contractible*, which would clearly be so if the action were linearizable.

[1] For a detailed exposition see the article [**Sch**] of SCHWARZ in this volume.

In this context the work [**KN78**] of KEMPF and NESS plays a fundamental role. They study the interaction between closed orbits of G and K-orbits and show the following (cf. [**Sch**], [**DK85**], [**PS85**]):

Theorem 3. *In the situation above there is a real algebraic subvariety $Y \subset X$ which is stable under the maximal compact subgroup $K \subset G$ and has the following properties:*

(a) *Y meets every closed G-orbit in X;*

(b) *For every $y \in Y$ the G-orbit Gy is closed and $Gy \cap Y = Ky$.*

As a consequence we see that the canonical map

$$\varphi : Y/K \longrightarrow X/\!\!/G$$

induced by the inclusion $Y \subset X$ is bijective, and one shows that φ is even a homeomorphism in the $\mathbf{C}$-topology. In particular, the quotient $X/\!\!/G$ carries the quotient topology (cf. Remark 4.1). We call the subvariety Y the *Kempf-Ness variety.*

The above results give a partial explanation of the fact that in many situations the quotient map $\pi : X \rightarrow X/\!\!/G$ behaves like a *proper* map. As an example, we have for every $y \in Y$ a canonical isomorphism

$$\varinjlim_{U \ni y} H^p(\pi^{-1}(U)) \simeq H^p(\pi^{-1}(y)),$$

where U runs through a system of neighbourhoods of y, which allows us to use in an efficient way the *Leray spectral sequence* of the quotient map π (cf. 7.3).

There is a strong relation of the KEMPF-NESS construction above with the *moment map* associated to the action of the compact group K. In fact, the variety $Y \subset X$ is the set of critical points of the moment map (see [**Ki84**], [**Ns84**]).

Another important result in this context is due to NEEMAN [**Ne85**]:

Theorem 4. *The Kempf-Ness variety $Y \subset X$ is a deformation retract of X. The deformation is a K-deformation and takes place along the closures of G-orbits.*

Using this result one obtains a short proof of Theorem 1 and 2 above (cf. [**Sch**]).

§ 5 Linearization

5.1. Linearizable actions. The first general result in connection with the Linearization Problem 0.2 is a consequence of LUNA's slice theorem 7.1 (cf. [**KP85**, Proposition 5.1]):

Theorem. *Let G be a reductive group acting on $\mathbf{C}^n$ and assume that every G-invariant function is a constant. Then the action is linearizable.*

It is even enough to assume that G is acting on a smooth *acyclic* affine variety X with only constant invariant functions. *Then X is G-isomorphic to a representation of G* (see 7.1). A typical application is when G has a dense orbit on such a variety.

5.2. In case of a torus action a more general result was obtained by BIALYNICKI-BIRULA [**BB67**].

Proposition. *If a torus T acts on $\mathbf{C}^n$ with an orbit of codimension ≤ 1, then the action is linearizable.*

This can be generalized to the case where one only assumes that the underlying affine variety X is smooth and acyclic and that the quotient $X/\!\!/T$ has dimension ≤ 1 (cf. Theorem 5.6).

5.3. The following result was obtained by KRAFT-POPOV [**KP85**] in dimension 3 and generalized to dimension 4 by PANYUSHEV [**Pa84**].

Theorem. *An action of a (connected) semisimple group G on $\mathbf{C}^n$ is linearizable for $n \leq 4$.*

The main idea in the proof is to study the quotient map $\pi : \mathbf{C}^n \to \mathbf{C}^n/\!\!/G$. The quotient $\mathbf{C}^n/\!\!/G$ is $\mathbf{C}^2$, $\mathbf{C}$ or a point, by our assumptions (see 4.2), and consequently π is flat. This means that the invariant ring is a polynomial ring in ≤ 2 variables and that the coordinate ring itself is a free G-module over the invariant ring.

5.4. The first open cases, which would also be needed for a generalization of Theorem 5.3 to (connected) reductive groups, are the $\mathbf{C}^*$-actions on $\mathbf{C}^3$ and $\mathbf{C}^4$.

Conjecture. *A $\mathbf{C}^*$-action on $\mathbf{C}^3$ is linearizable.*

We can prove this under the additional assumption that the quotient $\mathbf{C}^3/\!\!/\mathbf{C}^*$ is smooth (e.g. in case the fixed point set is positive dimensional).

A more general result has been obtained by KORAS and RUSSELL [**KR89**].

Remark. For finite groups acting on $\mathbf{C}^3$ almost nothing is known about the fixed point set or the Linearization Problem. A first interesting case should be an automorphism σ of order 2 with two-dimensional fixed point set $F = (\mathbf{C}^3)^\sigma$: Is F isomorphic to $\mathbf{C}^2$ or even ambient isomorphic (i.e. by an automorphism of $\mathbf{C}^n$) to a coordinate plane (cf. [**Je87**, Theorem 1.1])? Is the quotient $\mathbf{C}^3/\sigma$ isomorphic to $\mathbf{C}^3$? Is σ linearizable?

5.5. Fix-pointed actions. For actions with "many fixed points" we can prove a general decomposition theorem. Following BASS-HABOUSH we call an action of a reductive group G on an affine variety X *fix-pointed* if every closed orbit is a fixed point. This is equivalent to the condition that the quotient map $\pi : X \to X/\!\!/G$ induces an isomorphism $X^G \xrightarrow{\sim} X/\!\!/G$, i.e. that π has a G-equivariant section. The following theorem is a consequence of joint work with LUNA. It was obtained independently by BASS and HABOUSH [**BH85**]; their result is actually more general.

Theorem. *Let G be a reductive group acting on a smooth affine variety X. If the action is fix-pointed, then $\pi : X \to X/\!\!/G$ has the structure of a G-vector bundle (6.1).*

The Serre-conjecture (proved in [**Qu76**], [**Su76**]) implies the following (use 2.2 and 6.8):

Corollary. *For any fix-pointed action of a reductive group G on $\mathbf{C}^n$ there is a G-equivariant isomorphism $\mathbf{C}^n \xrightarrow{\sim} (\mathbf{C}^n)^G \times V$, where V is a G-module. In particular, the action is linearizable if* $\dim(\mathbf{C}^n)^G \leq 2$.

5.6. One-dimensional quotient. Generalizing 5.1 and 5.2, one might ask the following question:

> *Is any action of a reductive group G on $\mathbf{C}^n$ with a quotient $\mathbf{C}^n/\!\!/G$ of dimension ≤ 1 linearizable?*

LUNA outlined an attack on this problem in 1983, which has been the guide in the joint work [**KS89**] with SCHWARZ. There are several significant results which have been obtained up to now. The easiest to state is the following:

Theorem. *Let the reductive group G act on the smooth acyclic affine variety X. Suppose that* $\dim X/\!\!/G = 1$ *and that one of the following conditions holds:*

(a) *G acts semifreely on X (i.e. the generic orbit is closed and has trivial stabilizer).*

(b) *G is simple.*

(c) *G is a torus.*

Then X is G-isomorphic to a linear representation.

Nevertheless, the question has a negative answer in general as shown by the recent counterexamples due to SCHWARZ [**Sch89**]. On the other hand we can show (see [**KS89**]) that every such action is *holomorphically linearizable.*

5.7. Quadratic actions. A general linearization result was obtained by JURKIEWICS for actions which have a specific form in terms of the coordinate functions [**Ju88,Ju89**]. Let G be a reductive group and let $\varphi : G \to \mathcal{G}_n$ be an algebraic action on $\mathbf{C}^n$. If $0 \in \mathbf{C}^n$ is a fixed point then we can write φ in the form

$$\varphi(g) = \varphi_1(g) + \varphi_2(g) + \cdots + \varphi_s(g)$$

where $\varphi_i(g)$ is homogeneous of degree i (for all $g \in G$). Clearly, the map $g \mapsto \varphi_1(g)$ is a linear action and coincides with the tangent representation in 0.

Proposition (JURKIEWICZ). *Assume that the action φ has the form*

$$\varphi(g) = \varphi_1(g) + \varphi_d(g) + \cdots + \varphi_{2d-2}(g)$$

for some $d > 1$. Then φ is linearizable.

In particular, every action of the form $g \mapsto \varphi(g) = \lambda(g) + \theta(g)$ where $\lambda(g)$ is linear and $\theta(g)$ quadratic, is linearizable. More precisely, JURKIEWICZ shows that $\varphi(g) = \rho^{-1} \circ \varphi_1(g) \circ \rho$ for all $g \in G$, where $\rho := \int_G \varphi_1(g)^{-1} \circ \varphi(g)$. (Here $\int_G$ denotes the Reynolds operator, i.e., the canonical projection from the vector space of all polynomial endomorphisms of $\mathbf{C}^n$ to the subspace of all G-equivariant ones.)

§ 6 G-Vector Bundles

For a detailed exposition of this subject we refer to [**Kr89**] and to the forthcoming notes [**DMV89**].

6.1. Let G be an algebraic group acting on a variety X.

Definition. A *G-vector bundle* on X is a vector bundle $\mathcal{V}$ on X with an algebraic G-action such that the following holds:

(a) The projection $p : \mathcal{V} \to X$ is G-equivariant;

(b) The action is linear on the fibres $\mathcal{V}_x := p^{-1}(x)$, (i.e. for every $g \in G$ and $x \in X$ the map $v \mapsto gv : \mathcal{V}_x \to \mathcal{V}_{gx}$ is linear).

It follows from the definition that for every point $x \in X$ we obtain a *rational representation* of the isotropy group $G_x := \{g \in G \mid gx = x\}$ on the fibre $\mathcal{V}_x$.

Example. Every *G-module* M (i.e. finite dimensional rational representation of G) determines a G-vector bundle $\mathcal{V} := M \times X \xrightarrow{\text{pr}} X$. A G-vector bundle is called *trivial* if it is isomorphic to a G-vector bundle of this form.

The *categories of vector bundles* and of *G-vector bundles* on X will be denoted by $\text{Vec}(X)$ and $\text{Vec}_G(X)$, respectively.

Remark. If X is an *affine* variety with coordinate ring $\mathcal{O}(X)$, then the category of $\text{Vec}_G(X)$ is equivalent to the category $\text{Proj}_G(\mathcal{O}(X))$ of G-$\mathcal{O}(X)$-modules which are projective and of finite rank over $\mathcal{O}(X)$ and (locally finite and) rational as G-module. The functor is given by taking global sections: $\mathcal{V} \mapsto \mathcal{V}(X)$. It is easy to see that any $P \in \text{Proj}_G(\mathcal{O}(X))$ is also projective as a G-$\mathcal{O}(X)$-module, i.e. a direct summand of a free G-$\mathcal{O}(X)$-module $M \otimes_{\mathbb{C}} \mathcal{O}(X)$ (cf. [**BH85**, 4.2]).

6.2. Family of representations. If the G-action on X is trivial, a G-vector bundle can be understood as an *algebraic family of representations* $(\mathcal{V}_x)_{x \in X}$ of G. The following result is well known.

Proposition. *Let G be reductive acting trivially on X and let $\mathcal{V}$ be a G-vector bundle on X.*

(a) *$\mathcal{V}$ is locally trivial in the Zariski-topology. In particular, the representations $\mathcal{V}_x$ are all equivalent, if X is connected.*

(b) *There is an isomorphism of G-vector bundles*

$$\mathcal{V} \xrightarrow{\sim} \bigoplus_\omega M_\omega \otimes \mathcal{V}_\omega,$$

where each M_ω is a simple G-module and each $\mathcal{V}_\omega$ a vector bundle on X.

As a consequence, we see that every G-vector bundle on X is trivial in case every vector bundle on X is trivial.

6.3. Homogeneous Bundles. Now assume that X is a *homogeneous* G-variety, i.e. $X = G/H$ with a closed algebraic subgroup $H \subset G$. It is well known that every G-vector bundle on G/H is of the form

$$G \star^H N \to G/H.$$

Here N is an H-module and $G \star^H N$ is the orbit space $(G \times N)/H$ where the (right-)H-action is given by $(g, n) \cdot h := (gh, h^{-1} \cdot n)$. These bundles are usually called *homogeneous vector bundles.*

A G-vector bundle $G \star^H N$ is trivial if and only if the representation of H on N extends to a representation of G. As an example, we see that every $\mathrm{SL}_2(\mathbf{C})$-vector bundle on $\mathbf{C}^2 \setminus \{0\}$ is trivial. In fact, $\mathbf{C}^2 \setminus \{0\} \xleftarrow{\sim} \mathrm{SL}_2(\mathbf{C})/U$ where $U = \left\{ \begin{pmatrix} 1 & * \\ 0 & 1 \end{pmatrix} \right\} \xrightarrow{\sim} \mathbf{C}^+$, and every representation of U extends to a representation of $\mathrm{SL}_2(\mathbf{C})$ (cf. [**Kr84**, Lemma III.3.9]).

6.4. Principal bundles. Let us consider a principal G-bundle $\pi : X \to Y$. This means that X is a G-variety with all orbits isomorphic to G, that the fibres of π are G-orbits and that π is locally trivial in the *étale topology*, i.e. there is a surjective étale morphism $Y' \to Y$ and a G-isomorphism

$$\begin{array}{ccc} G \times Y' & \xrightarrow{\sim} & X \times_Y Y' \\ \downarrow & & \downarrow \\ Y' & = & Y' \end{array}$$

over Y'.

Proposition. *Let $\pi : X \to Y$ be a principal G-bundle. The pull back $\mathcal{W} \mapsto \pi^*\mathcal{W}$ defines an equivalence:*

$$\pi^* : \mathrm{Vec}(Y) \xrightarrow{\sim} \mathrm{Vec}_G(X).$$

Again this is well known and can be proved in several ways.

6.5. Pull-backs. Now assume that G is reductive acting on an affine variety X, and denote by $\pi : X \to X/\!\!/G$ the quotient (4.1). If $\mathcal{W}$ is a vector bundle on $X/\!\!/G$ then the pullback $\pi^*\mathcal{W}$ is clearly a G-vector bundle on X, and we obtain the following diagram:

$$\begin{array}{ccc} \pi^*\mathcal{W} & \longrightarrow & X \\ \downarrow \tilde{\pi} & & \downarrow \pi \\ \mathcal{W} & \longrightarrow & X/\!\!/G \end{array}$$

It follows that $\tilde{\pi}$ is a quotient map, i.e. $\mathcal{W} \simeq \pi^*\mathcal{W}/\!\!/G$ as a vector bundle over $X/\!\!/G$. The following proposition characterizes the G-vector bundles which are obtained in this way.

Proposition. *A G-vector bundle $\mathcal{V}$ on X is isomorphic to a pullback $\pi^*\mathcal{W}$ if and only if it satisfies the following condition:*

(PB) *For every $x \in X$ such that the orbit Gx is closed, the isotropy group G_x acts trivially on the fibre $\mathcal{V}_x$.*

In this case $\mathcal{V}/\!\!/G$ is a vector bundle over $X/\!\!/G$ isomorphic to $\mathcal{W}$.

Corollary. *Assume that the isotropy groups G_x generate G. Then a vector bundle $\mathcal{W}$ on $X/\!\!/G$ is trivial if and only if $\pi^*\mathcal{W}$ is a trivial G-vector bundle.*

(If $\pi^*\mathcal{W} \simeq M \times X$, then every isotropy group G_x acts trivially on M. Now it follows from the assumption that M is the trivial G-module, hence $\mathcal{W} \simeq \pi^*\mathcal{W}/\!\!/G \simeq M \times X/\!\!/G$ is trivial, too.)

6.6. Linearization. There is the following interesting connection between G-vector bundles and the Linearization Problem 0.2 (see [**Kr89**]).

Proposition. *Assume that the* Linearization Problem *has a positive answer, and let V be a representation of the reductive group G. Then*

(a) *Every G-vector bundle on V is trivial.*

(b) *Every vector bundle on $V/\!\!/G$ is trivial.*

The counterexamples of SCHWARZ [**Sch89**] to the Linearization Problem are non-trivial G-vector bundles over representations. In particular, the equivariant Serre-conjecture stating that *every G-vector bundle over a representation space V of G is trivial*, is false in general. It is easy to see that it is true for *line bundles* (see [**Kr89**, § 7]). It also holds in the differentiable setting, since V has an obvious G-retraction to the origin.

6.7. Stability. A G-vector bundle $\mathcal{V}$ on a G-variety X is called *stably trivial* if there is a trivial G-vector bundle $\mathcal{V}_0 = M \times X$ such that $\mathcal{V} \oplus \mathcal{V}_0$ is trivial. The following result is due to BASS-HABOUSH [**BH87**] and THOMASON.

Theorem. *Every G-vector bundle on a representation space V is stably trivial.*

In fact, their results are more general and are expressed in terms of algebraic K-Theory. In view of the proposition above we may ask the following question:

Problem. *Is it true that every vector bundle on $V/\!\!/G$ is stably trivial?*

6.8. Some results. For *fix-pointed* actions (5.5) we have the following result due to Bass and Haboush [**BH87**, 10.2]:

Theorem 1. *Let X be a fix-pointed G-variety with quotient $\pi : X \to X/\!\!/G$. Then every G-vector bundle on X is of the form*

$$\bigoplus_\omega M_\omega \otimes \pi^* \mathcal{V}_\omega$$

where each M_ω is a simple G-module and each $\mathcal{V}_\omega$ a vector bundle on $X/\!\!/G$.

Under the assumptions of the theorem the quotient morphism $\pi : X \to X/\!\!/G$ induces an isomorphism $X^G \xrightarrow{\sim} X/\!\!/G$ (5.5). In particular, we obtain the following corollary:

Corollary. *Let X be as above and assume that every vector bundle on $X/\!\!/G$ or on X is trivial. Then every G-vector bundle on X is trivial.*

Typical examples for the corollary are:

(a) Fix-pointed actions on $\mathbf{C}^n$;

(b) Actions on affine varieties with a fixed point and without non-constant invariant functions.

In particular, we see that for a representation V without invariants every G-vector bundle is trivial. This is generalized in the following theorem:

Theorem 2. *Let V be a representation of G. Assume that G acts semi-freely (i.e. the generic orbit is closed and has a trivial stabilizer) and that* $\dim V/\!\!/G \leq 1$. *Then every G-vector bundle on V is trivial.*

This can be proved using methods developed in [**KS89**], where we study the Linearization Problem for actions with one-dimensional quotients (see 5.6).

The last result here deals with tori. For an outline of the proof we refer to [**Kr89**].

Theorem 3. *Let V be a representation of a torus T. Assume that the principal stratum $(V/\!\!/T)_{\mathrm{pr}}$ (see 7.2) has a complement of codimension ≥ 2 and that the quotient $V/\!\!/T$ is factorial. Then $V/\!\!/T$ is smooth, $\pi : V \to V/\!\!/T$ has a section and every T-vector bundle on V is trivial.*

The first examples are given by the representations of $\mathbf{C}^*$ on $\mathbf{C}^3$ with weights 1, n and $-m$, where n and m are positive numbers.

§ 7 Slices and Stratification

7.1. One of the most important results for the study of reductive transformation groups and quotient maps is LUNA's slice theorem [**Lu73**] (cf. [**Lu75a**] or [**DMV89**]).

Slice Theorem. *Consider a reductive group G acting on an affine variety X. Given a closed orbit O and a point $x \in O$ with stabilizer H, there is a locally closed H-stable affine subvariety S containing x such that the diagram*

$$\begin{array}{ccc} G\star^H S & \xrightarrow{\varphi} & X \\ \pi\downarrow & & \pi\downarrow \\ S/\!\!/H & \xrightarrow{\overline{\varphi}} & X/\!\!/G \end{array}$$

is a fibre product where φ is G-equivariant and $\varphi, \overline{\varphi}$ are étale maps. If X is smooth at x, then S is locally isomorphic to the slice representation of H at x.

Here $G \star^H S$ denotes the *associated bundle* to the principle H-bundle $G \to G/H$ (see 6.3). The *slice representation* of H at x is the representation of H on the normal space $N_x := T_x(X)/T_x(O)$. By a result due to MATSUSHIMA the stabilizer of a point of a closed orbit is always *reductive*; in particular, we have an H-stable decomposition $T_x(X) \simeq N_x \oplus T_x(O)$. Finally S *locally isomorphic to the slice representation* N_x means that there is an étale G-equivariant morphism $\psi : S \to N_x$ sending x to $0 \in N_x$ such that the diagram

$$\begin{array}{ccc} S & \xrightarrow{\psi} & N_x \\ \pi\downarrow & & \pi\downarrow \\ S/\!\!/H & \xrightarrow{\overline{\psi}} & N_x/\!\!/H \end{array}$$

is a fibre product.

Let us remark that Theorem 5.1 is an easy consequence of the Slice Theorem. In fact, if $X/\!\!/G$ is a point, then X is isomorphic to $G\star^H N$ where N is a representation H; hence it is a vector bundle over G/H. But then X is acyclic only when $G = H$.

7.2. Stratification of quotients. The slice representation of a smooth affine G-variety X defines an equivalence relation on the quotient $X/\!\!/G$: Two points $y_1, y_2 \in X/\!\!/G$ are *equivalent* if the slice representations in $x_1 \in O_{y_1}$ and $x_2 \in O_{y_2}$ are conjugate. (Remember that O_y denotes the unique closed orbit in the fibre $\pi^{-1}(y)$.) As a consequence of the slice theorem this defines a *finite stratification* $X/\!\!/G = \bigcup_\lambda (X/\!\!/G)_\lambda$ into locally closed subvarieties with the property that the induced maps

$$\pi_\lambda : X_\lambda := \pi^{-1}((X/\!\!/G)_\lambda) \longrightarrow (X/\!\!/G)_\lambda$$

are *G-fibrations* (in the étale topology; cf. [**Lu73**]). In particular, all strata $(X/\!\!/G)_\lambda$ are *smooth* subvarieties of $X/\!\!/G$. The open stratum is usually called the *principal stratum.*

This result can be used to prove Corollary 5.5 about fix-pointed actions on $\mathbf{C}^n$: In this case there is only one stratum and so $\pi : \mathbf{C}^n \to \mathbf{C}^n/\!\!/G$ is a G-fibre bundle, whose structure group is the group A of all G-automorphisms of the fibre W (which is a G-module). It is easy to see that A is a semi-direct product of a unipotent group U by a product of GL_n's, and it is well known that every principal bundle over $\mathbf{C}^n$ with such a structure group is trivial (use the Serre conjecture).

7.3. Another consequence of the Slice Theorem and the induced stratification is the following. Let k be any field and denote by $\boldsymbol{k}$ the constant sheaf with stalk k on the smooth affine G-variety X. *Then the higher direct images* $\mathrm{R}^i\pi_*\boldsymbol{k}$ *are locally constant on the strata* $(X/\!\!/G)_\lambda$ *with stalks isomorphic to the cohomology group* $\mathrm{H}^i(O_\lambda, k)$, *where* O_λ *is the closed orbit in a fibre over* $(X/\!\!/G)_\lambda$. This can be used to prove the following:

Theorem. *Let G be a reductive group acting on a smooth contractible affine variety X with one-dimensional quotient. Then $X/\!\!/G \simeq \mathbf{C}$, and either the action is fix-pointed and X is isomorphic to a G-module, or there is exactly one fixed point $x_0 \in X$ and two strata, namely $\{\pi(x_0)\}$ and $\mathbf{C} \setminus \{\pi(x_0)\}$.*

In particular, we see that the direct images $\mathrm{R}^i\pi_*\boldsymbol{k}$ are completely determined by the *generic stalk* $\mathrm{H}^i(O, k)$ where O is the generic closed orbit in X, and the *monodromy* $\alpha \in \mathrm{Aut}_k(\mathrm{H}^i(O, k))$. If X is a G-module V with invariant ring generated by a homogeneous function p of degree n, the monodromy is always of finite order d dividing n. What is the meaning of the monodromy α in terms of the representation?

References

[**AM75**] Abhyankar, S.S; Moh, T.-T.: *Embeddings of the line in the plane.* J. Reine Angew. Math. **276** (1975), 149–166

[**Ba84**] Bass, H.: *A non-triangular action of G_a on $\mathbf{A}^3$.* J. Pure Appl. Algebra **33** (1984), 1–5

[Ba85] Bass, H.: *Algebraic group actions on affine spaces.* In: Group actions on rings. Proc. AMS-IMS-SIAM Summer Res. Conf., Brunswick/Maine 1984. Contemp. Math. **43** (1985), 1–23

[BCW82] Bass, H.; Connell, E. H.; Wright, D.: *The Jacobian conjecture: Reduction of degree and formal expansion of the inverse.* Bull. Amer. Math. Soc. **7** (1982), 287–330

[BH85] Bass, H.; Haboush, W.: *Linearizing certain reductive group actions.* Trans. Amer. Math. Soc. **292** (1985), 463–482

[BH87] Bass, H.; Haboush, W.: *Some equivariant K-theory of affine algebraic group actions.* Comm. Algebra **15** (1987), 181–217

[BC85] Beauville, A.; Colliot-Thélène, J.-L.; Sansuc, J.-J.; Swinnerton-Dyer, P.: *Variétés stablement rationnelles non rationnelles.* Ann. of Math. **121** (1985), 283–318

[BB67] Bialynicki-Birula, A.: *Remarks on the action of an algebraic torus on k^n, I and II.* Bull. Acad. Pol. Sci **14** (1966), 177–181 and **15** (1967), 123–125

[Bt87] Boutot, J.-F.: *Singularités rationnelles et quotients par les groupes réductifs.* Invent. math. **88** (1987), 65–68

[Br72] Bredon, G. E.: *Introduction to Compact Transformation Groups.* Pure and Applied Mathematics vol. **46**, Academic Press 1972

[DK85] Dadok, J.; Kac, V.: *Polar representations.* J. Algebra **92** (1985), 504–524

[DMV89] Kraft, H.; Slodowy, P.; Springer, A. T.: *Algebraische Transformationsgruppen und Invariantentheorie.* DMV-Seminar Notes; to appear

[Da89] Danielewski, W.: *On the cancellation problem and automorphism group of affine algebraic varieties.* Preprint 1989

[DP] tom Dieck, T.; Petrie, T.: *Homology planes: An announcement and survey.* In this volume

[Fo73] Fogarty, J.: *Fixed point schemes.* Amer. J. Math **95** (1973), 35–51

[Fu79] Fujita, T.: *On Zariski problem.* Proc. Japan Acad. Ser. A Math. Sci. **55** (1979), 106–110

[Fr82] Furushima, M.: *Finite groups of polynomial automorphisms in the complex affine plane (I).* Mem. Fac. Sci. Kyushu Univ. **36** (1982), 85–105

[GD75] Gizatullin, M. H.; Danilov, V. I.: *Automorphisms of affine surfaces, I and II.* Math. USSR-Izv. **9** (1975), 493–534, and Math. USSR-Izv. **11** (1977), 51–98

[Gu62] Gutwirth, A.: *The action of an algebraic torus on the affine plane.* Trans. Amer. Math. Soc. **105** (1962), 407–414

[HS82] Hsiang, W.-Y.; Straume, E.: *Actions of compact connected Lie groups with few orbit types.* J. Reine Angew. Math. **334** (1982), 1–26

[**HS86**] Hsiang, W.-Y.; Straume, E.: *Actions of compact connected Lie groups on acyclic manifolds with low dimensional orbit space.* J. Reine Angew. Math. **369** (1986), 21–39

[**Ig77**] Igarashi, T.: *Finite subgroups of the automorphism group of the affine plane.* M.A. thesis, Osaka University (1977),

[**Je87**] Jelonek, Z.: *The extension of regular and rational embeddings.* Math. Ann. **113** (1987), 113–120

[**Ju88**] Jurkiewicz, J.: *On the linearization of actions of linearly reductive groups.* Comment. Math. Helv.,to appear

[**Ju89**] Jurkiewicz, J.: *On some reductive group actions on affine space.* Preprint 1989

[**Ka79**] Kambayashi, T.: *Automorphism group of a polynomial ring and algebraic group actions on an affine space.* J. Algebra **60** (1979), 439–451

[**Ka80**] Kambayashi, T.: *On Fujita's strong cancellation theorem for the affine space.* J. Fac. Sci. Univ. Tokyo **23** (1980), 535–548

[**KR82**] Kambayashi, T.; Russel, P.: *On linearizing algebraic torus actions.* J. Pure Appl. Algebra **23** (1982), 243–250

[**Ke80**] Kempf, G.: *Some quotient surfaces are smooth.* Michigan Math. J. **27** (1980), 295–299

[**KN78**] Kempf, G.; Ness, L.: *The length of a vector in a representation space.* In: Algebraic Geometry. Lecture Notes in Mathematics **732**, Springer-Verlag 1978, 233–244

[**Ki84**] Kirwan, F.: *Cohomology of quotients in symplectic and algebraic geometry.* Mathematical Notes **31**, Princeton University Press 1984

[**KR89**] Koras M.; Russell, P.: *On linearizing "good" $\mathbf{C}^3$-actions on $\mathbf{C}^3$.* Proceedings of the Conference on Group Actions and Invariant Theory, Montreal 1988

[**Kr84**] Kraft, H.: *Geometrische Methoden in der Invariantentheorie.* Aspekte der Mathematik **D1**, Vieweg-Verlag 1984

[**Kr85**] Kraft, H.: *Algebraic group actions on affine spaces.* In: Geometry today. Progress in Mathematics **60**, Birkhäuser Verlag 1985, 251–265

[**Kr89**] Kraft, H.: *G-vector bundles and the linearization problem.* Proceedings of the Conference on Group Actions and Invariant Theory, Montreal 1988

[**KP85**] Kraft, H.; Popov, V. L.: *Semisimple group actions on the three dimensional affine space are linear.* Comment. Math. Helv. **60** (1985), 466–479

[**KPR86**] Kraft, H.; Petrie, T.; Randall, J.: *Quotient varieties.* Preprint 1986. To appear in Adv. math.

[**KS89**] Kraft, H.; Schwarz, G.: *Linearizing reductive group actions on affine space with one-dimensional quotient.* Proceedings of the Conference on Group Actions and Invariant Theory, Montreal 1988

[Ku53] van der Kulk, W.: *On polynomial rings in two variables.* Nieuw Arch. Wisk. **1** (1953), 33–41

[Lu73] Luna, D.: *Slices étales.* Bull. Soc. Math. France, Mémoire **33** (1973), 81–105

[Lu75a] Luna, D.: *Adhérences d'orbite et invariants.* Invent. Math. **29** (1975), 231–238

[Lu75b] Luna, D.: *Sur certaines opérations différentiables des groupes de Lie.* Amer. J. Math. **97** (1975), 172–181

[Lu76] Luna, D.: *Fonctions différentiables invariantes sous l'operation d'un groupe réductif.* Ann. Inst. Fourier **26** (1976), 33–49

[Mi71] Miyanishi, M.: *G_a-actions of the affine plane.* Nagoya Math. J. **41** (1971), 97–100

[Mi84] Miyanishi, M.: *An algebro-topological characterization of the affine space of dimension three.* Amer. J. Math. **106** (1984), 1469–1485

[MS80] Miyanishi, M.; Sugie, T.: *Affine surfaces containing cylinderlike open sets.* J. Math. Kyoto Univ. **20** (1980), 11–42

[MF82] Mumford, D.; Fogarty, J.: *Geometric Invariant Theory.* 2nd edition. Ergeb. Math. und Grenzgeb. **34**, Springer-Verlag 1982

[Na72] Nagata, M.: *On automorphism group of* $k[\mathbf{x},\mathbf{y}]$. Lectures in Mathematics **5** (1972),, Kyoto University

[Ne85] Neeman, A.: *The topology of quotient varieties.* Ann. Math. **122** (1985), 419–459

[Ns84] Ness, L.: *A stratification of the nullcone via the moment map.* Amer. J. Math. **106** (1984), 1281–1330

[Ol76] Oliver, R.: *A proof of the Conner conjecture.* Ann. Math. **103** (1976), 637–644

[Ol79] Oliver, R.: *Weight systems for* SO(3)*-actions.* Ann. of Math. **110** (1979), 227–241

[Pa84] Panyushev, D. I.: *Semisimple automorphism groups of four-dimensional affine space.* Math. USSR-Izv. **23** (1984), 171–183

[PR86] Petrie, T.; Randall, J. D.: *Finite-order algebraic automorphisms of affine varieties.* Comment. Math. Helv. **61** (1986), 203–221

[Po87] Popov, V. L.: *On actions of G_a on $\mathbf{A}^n$.* In: Algebraic Groups (Utrecht 1986), Lecture Notes in Mathematics **1271**, Springer-Verlag 1987

[PS85] Procesi, C.; Schwarz, G.: *Inequalities defining orbit spaces.* Invent. math. **81** (1985), 539–554

[Qu76] Quillen, D.: *Projective modules over polynomial rings.* Invent. math. **36** (1976), 167–171

[Ra71] Ramanujam, C. P.: *A topological characterization of the affine plane as an algebraic variety.* Ann. Math **94** (1971), 69–88

[**Re68**] Rentschler, R.: *Opérations du groupe additif sur le plan affine.* C. R. Acad. Sci. Paris **267 A** (1968), 384–387

[**Ru81**] Russel, P.: *On affine-ruled rational surfaces.* Math. Ann. **255** (1981), 287–302

[**Sch**] Schwarz, G.: *The topology of algebraic quotients.* In this volume

[**Sch89**] Schwarz, G.: *Exotic algebraic group actions.* Preprint 1989. To appear in C. R. Acad. Sci. Paris

[**Se80**] Serre, J. P.: *Trees.* Springer-Verlag 1980

[**Sh66**] Shafarevich, I. R.: *On some infinite dimensional groups.* Rend. Math. e Appl. (5) **25** (1966), 208–212

[**Sh**] Shepherd-Barron, N.: *Rationality of moduli spaces via invariant theory.* In this volume

[**Sn**] Snow, D.: *Unipotent actions on affine space.* In this volume

[**Su**] Sugie, T.: *Algebraic characterization of the affine plane and the affine 3-space.* In this volume

[**Su76**] Suslin, A.: *Projective modules over a polynomial ring.* Dokl. Akad. Nauk SSSR **26** (1976), (in Russian)

[**Ve73**] Verdier, J.-L.: *Caracteristique d'Euler-Poincaré.* Bull. Soc. math. France **101** (1973), 441–445

Hanspeter Kraft
Mathematisches Institut
Universität Basel
Rheinsprung 21
CH-4051 Basel
SWITZERLAND

ALMOST HOMOGENEOUS ARTIN–MOIŠEZON VARIETIES UNDER THE ACTION OF $PSL_2(C)$

DOMINGO LUNA, LUCY MOSER-JAUSLIN AND THIERRY VUST

Let X be an analytic compact connected complex variety. Denote by $\mathcal{M}(X)$ the field of global meromorphic functions on X. About 1950 it was shown that the transcendence degree of $\mathcal{M}(X)$ over $\mathbf{C}$ is less than or equal to the dimension of X ([**Si**], [**GA**] p.63). Those varieties for which equality holds we call Artin-Moišezon varieties; they were extensively studied around the end of the 1960's ([**M**], [**A**], ...). They have an "algebraic structure" which generalizes the notion of algebraic structure on classical algebraic varieties (i.e. proper smooth schemes of finite type over $\mathbf{C}$). Although at one time it was not clear that nonprojective Artin-Moišezon varieties existed ([**CK**], [**Hi**] Problem 34, [**Nag**]), there are by now many examples ([**Ha**] Appendix B, [**M-S**]), and the term "algebraic variety" is often used synonymously with Artin-Moišezon variety.

This article is a contribution to the study of the actions of complex semisimple groups on Artin-Moišezon varieties. In §1 we construct examples of (smooth compact connected) three-dimensional Artin-Moišezon varieties which are not schemes and on which $PSL_2(\mathbf{C})$ acts with an open (and dense) orbit. In §2 we verify some of the points that were left undone in §1. We show in §3, using only the local structure of the orbits of $PSL_2(\mathbf{C})$, that the varieties constructed in §1 are not schemes. This proof shows that the behavior of semisimple complex group actions can be quite different on Artin-Moišezon varieties than on classical algebraic varieties. Finally, in §4 we make some closing comments and pose questions suggested by this construction.

§1. A construction of examples

Let Γ be a non-trivial finite subgroup of the group $G = \mathrm{PSL}_2(\mathbf{C})$ (i.e. Γ can be cyclic, dihedral, octahedral or icosahedral). In this section, we describe a procedure to construct (smooth, compact, connected) Artin-Moišezon varieties X such that:

- G acts (morphically) on X with an open and dense orbit isomorphic to G/Γ;
- X is not a scheme.

We denote by M the vector space of 2×2 matrices with complex coefficients, and by $\mathbf{P}(M) \simeq \mathbf{P}_3$ the corresponding projective space. The group $G \times G$ acts in a natural way on $\mathbf{P}(M)$ (by left and right multiplication). Let B be a Borel subgroup of G, and let T be a maximal torus of B. We denote by ${}^B\mathbf{P}(M)$ the set of elements of $\mathbf{P}(M)$ stable by the left action of B; clearly we have that ${}^B\mathbf{P}(M) \simeq \mathbf{P}_1$. There are two types of orbits of G acting on $\mathbf{P}(M)$ by left multiplication:

- one open orbit isomorphic to G;
- the orbits $G \cdot z$, $z \in {}^B\mathbf{P}(M)$, each isomorphic to $\mathbf{P}_1$.

The group Γ acts by right multiplication on ${}^B\mathbf{P}(M)$. Since this action is faithful, Γ has orbits in ${}^B\mathbf{P}(M)$ consisting of $r = \mathrm{card}(\Gamma)$ elements. Choose such an orbit $\{z_1, \dots, z_r\}$, and set $Z_i = G \cdot z_i$ $(i = 1, \dots, r)$.

We denote by $X^{(1)}$ the (projective smooth) variety obtained from $\mathbf{P}(M)$ by blowing up the curves $Z_1, \dots, Z_r$. The action of $G \times \Gamma$ on $\mathbf{P}(M)$ extends to $X^{(1)}$ in such a way that the natural morphism $\pi^{(1)}: X^{(1)} \to \mathbf{P}(M)$ is equivariant. We will see in §2 that $(\pi^{(1)})^{-1}(Z_i)$ consists of two G-orbits, one (denoted $Z_i^{(1)}$) isomorphic to $\mathbf{P}_1$, and the other (denoted $Y^{(1)}$) isomorphic to G/T.

We construct the (projective smooth) variety $X^{(2)}$ by blowing up the curves $Z_1^{(1)}, \dots, Z_r^{(1)}$ of $X^{(1)}$. As before, $G \times \Gamma$ acts on $X^{(2)}$, and the natural morphism $\pi^{(2)}: X^{(2)} \to X^{(1)}$ is equivariant.

We set $Y_i^{(2)} = (\pi^{(2)})^{-1}(Y_i^{(1)}) \xrightarrow{\sim} Y_i^{(1)}$ $(i = 1, \dots, r)$. We will see in §2 that $\overline{Y_i^{(2)}}$ (the closure of $Y_i^{(2)}$ in $X^{(2)}$) consists of two G-orbits, $Y_i^{(2)}$ and a closed orbit (denoted $Z_i^{(2)}$) isomorphic to $\mathbf{P}_1$.

We denote by $X^{(3)}$ the (projective smooth) variety obtained by blowing up the curves $Z_1^{(2)}, \dots, Z_r^{(2)}$ in $X^{(2)}$. As in the previous constructions $G \times \Gamma$ acts on $X^{(3)}$, and the natural morphism $\pi^{(3)}: X^{(3)} \to X^{(2)}$ is equivariant.

Set $Y_i^{(3)} = (\pi^{(3)})^{-1}(Y_i^{(2)}) \xrightarrow{\sim} Y_i^{(2)}$, and denote by $\overline{Y_i^{(3)}}$ its closure in $X^{(3)}$ $(i = 1, \ldots, r)$.

Since $Y_i^{(3)} \simeq G/T$, there are at most two G-morphisms $\overline{Y_i^{(3)}} \to \mathbf{P}_1 \simeq G/B$. We will see in the next section that each $\overline{Y_i^{(3)}}$ in $X^{(3)}$ can be blown down to $\mathbf{P}_1$ via one of these two morphisms to obtain a smooth projective variety. Since the $\overline{Y_i^{(3)}}$'s $(i = 1, \ldots, r)$ are disjoint, one can perform all of these blowing downs simultaneously to obtain a smooth scheme $X^{(4)}$ on which $G \times \Gamma$ acts (we will see at the end of §3 that this scheme is not projective!). Denote by $\pi^{(4)}: X^{(3)} \to X^{(4)}$ the natural morphism, and set $Z_i^{(4)} = \pi^{(4)}\left(\overline{Y_i^{(3)}}\right) \simeq \mathbf{P}_1 \quad (i = 1, \ldots, r)$.

We set $X^{(5)} = X^{(4)}/\Gamma$, the quotient in the category of Artin-Moišezon spaces (see [C]). The group G acts on $X^{(5)}$, and the image of $Z_1^{(4)} \cup \ldots \cup Z_r^{(4)}$ in $X^{(5)}$ (denoted $Z^{(5)}$) is a G-orbit isomorphic to $\mathbf{P}_1$. Since Γ acts freely in an open neighborhood of $Z_1^{(4)} \cup \ldots \cup Z_r^{(4)}$, it is clear that $Z^{(5)}$ consists of smooth points in $X^{(5)}$.

$$\mathbf{P}(M) \xleftarrow{\pi^{(1)}} X^{(1)} \xleftarrow{\pi^{(2)}} X^{(2)} \xleftarrow{\pi^{(3)}} X^{(3)} \xrightarrow{\pi^{(4)}} X^{(4)} \xrightarrow{\pi^{(5)}} X^{(5)}$$

Since the Artin-Moišezon space $X^{(5)}$ is normal, and since G does not have any fixed points in $X^{(5)}$, the set of singular points of $X^{(5)}$ is a finite union of one-dimensional orbits (each isomorphic to $\mathbf{P}_1$). By a composition of blowing ups of such orbits, we obtain a (smooth) Artin-Moišezon variety X on which G acts, and the G-morphism $\pi: X \to X^{(5)}$ is an isomorphism in an open neighborhood of $Z = \pi^{-1}(Z^{(5)})$. By construction, G has an open orbit in X isomorphic to G/Γ. We will explain in §3 why X is not a scheme.

§2. Verifications needed for the construction of §1

We keep the notation of §1. Remember that $Z_i \ (i = 1, \ldots, r)$ are certain closed orbits of G in $\mathbf{P}(M)$. We set $\mathbf{P}(M)_i = \mathbf{P}(M) \setminus \bigcup_{j \neq i} Z_i$: it is a G-stable open subvariety containing Z_i. Since the items we must verify are all local, it is enough to consider the open subvarieties of $X^{(j)} \ (j = 1, \ldots, r)$ lying over $\mathbf{P}(M)_i$; these, we will see, can be considered as locally trivial fibre bundles over G/B_i for a special choice of a Borel subgroup B_i of G, such that the fibration is equivariant. Since the fibre is two-dimensional, we can then apply the classical theory of surfaces to obtain the desired results. We set $X_i^{(1)} = (\pi^{(1)})^{-1}(\mathbf{P}(M)_i)$. We start by analyzing the G-morphism $\pi^{(1)}: X_i^{(1)} \to \mathbf{P}(M)_i$.

First we fix some notation. Given H an algebraic subgroup of G and S an H-variety, we denote by $G \star^H S$ the G-variety obtained as the quotient of $G \times S$ by the action of H defined by $h \cdot (g, s) = (gh^{-1}, h \cdot s)$ $(h \in H, g \in G, s \in S)$.

Each Z_i has a unique point d_i corresponding to a line $\mathbf{C}\delta_i$ in M such that δ_i is nilpotent. We denote by B_i the Borel subgroup of G which fixes d_i. Set $N_i = \{\alpha \in M \mid \alpha\delta_i \in \mathbf{C}\delta_i\}$: it is a three-dimensional vector subspace of M. The hyperplane $\mathbf{P}(N_i) \subset \mathbf{P}(M)$ is stable by B_i. There are three kinds of orbits of B_i in $\mathbf{P}(N_i)$:

- one open orbit isomorphic to B_i;
- the orbit $Z_i \setminus \{d_i\}$ isomorphic to B_i/T_i, where T_i is a maximal torus in B_i;
- the fixed points ${}^{B_i}\mathbf{P}(M) \subset \mathbf{P}(N_i)$.

It follows that the natural morphism $G \star^{B_i} \mathbf{P}(N_i) \to \mathbf{P}(M)$ is birational and proper, that it is an isomorphism above $\mathbf{P}(M) \setminus Z_i$, and that it contracts $G \star^{B_i} Z_i \subset G \star^{B_i} \mathbf{P}(N_i)$ onto Z_i. Set $S_i^{(1)} = \mathbf{P}(N_i) \cap \mathbf{P}(M)_i$. From the remarks above we see that $G \star^{B_i} S_i^{(1)} \to \mathbf{P}(M)_i$ is exactly $\pi^{(1)}: X_i^{(1)} \to \mathbf{P}(M)_i$. In particular, it follows that $(\pi^{(1)})^{-1}(Z_i)$ consists of two orbits: $Z^{(1)} = G \star^{B_i} \{d_i\} \simeq \mathbf{P}_1$ and $Y_i^{(1)} = G \star^{B_i} (Z_i \setminus \{d_i\}) \simeq G/T$, as stated in §1.

We denote by $S_i^{(2)}$ the surface obtained by blowing up d_i in $S_i^{(1)}$. The group B_i acts on $S_i^{(2)}$. Denote by $C_i^{(2)}$ the strict transform of Z_i in $S_i^{(2)}$. The curve $C_i^{(2)}$ is isomorphic to $\mathbf{P}_1$, and the group B_i has two orbits in $C_i^{(2)}$: a fixed point (denoted $d_i^{(2)}$) and the orbit $C_i^{(2)} \setminus \{d_i\} \xrightarrow{\sim} Z_i \setminus \{d_i\} \simeq \mathbf{C}$. We denote by $S_i^{(3)}$ the surface obtained by blowing up $d_i^{(2)}$ in $S_i^{(2)}$ and by $C_i^{(3)}$ the strict transform of $C_i^{(2)}$ in $S_i^{(3)}$. The group B_i acts on $S_i^{(3)}$, and $C_i^{(3)}$ is stable under this action.

Set $X_i^{(2)} = (\pi^{(2)})^{-1}\left(X_i^{(1)}\right)$ and $X_i^{(3)} = (\pi^{(3)})^{-1}\left(X_i^{(2)}\right)$. It is clear that $\pi^{(2)}: X_i^{(2)} \to X_i^{(1)}$ is exactly $G \star^{B_i} S_i^{(2)} \to G \star^{B_i} S_i^{(1)}$ and $\pi^{(3)}: X_i^{(3)} \to X_i^{(2)}$ is exactly $G \star^{B_i} S_i^{(3)} \to G \star^{B_i} S_i^{(2)}$. In particular, it follows that $\overline{Y_i^{(2)}}$ consists of two orbits $Y_i^{(2)}$ and $Z_i^{(2)} = G \star^{B_i} \{d_i^{(2)}\} \simeq \mathbf{P}_1$, as stated in §1.

The curve Z_i has self-intersection 1 in $\mathbf{P}(N_i)$. Thus the self-intersection of $C_i^{(2)}$ in $S_i^{(2)}$ is 0, and the self-intersection of $C_i^{(3)}$ in $S_i^{(3)}$ is -1. Since $C_i^{(3)}$ is isomorphic to $\mathbf{P}_1$, the criterion of Castelnuovo implies that $C_i^{(3)}$ can be blown down in $S_i^{(3)}$ to obtain a smooth surface $S_i^{(4)}$. Denote $d_i^{(4)}$ the image of $C_i^{(3)}$ in $S_i^{(4)}$. The morphism $X_i^{(3)} = G \star^{B_i} S_i^{(3)} \to G \star^{B_i} S_i^{(4)}$ is the contraction of $\overline{Y_i^{(3)}} = G \star^{B_i} C_i^{(3)}$ onto $G \star^{B_i} \{d_i^{(4)}\} \simeq \mathbf{P}_1$. Thus, as stated on §1, one can contract each $\overline{Y_i^{(3)}}$ in $X^{(3)}$ to $\mathbf{P}_1$ to obtain a projective smooth G-variety.

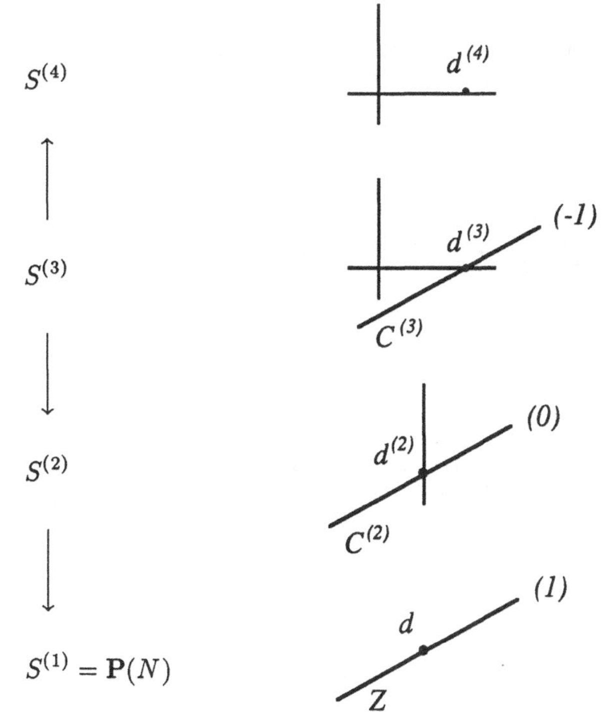

We have finished the verification of the details needed for §1. What follows will be used in the next section.

Denote by e the point in $\mathbf{P}(M)$ corresponding to the identity matrix in M. Let $e^{(3)}$ be the point of $X^{(3)}$ such that $\pi^{(1)}\left(\pi^{(2)}\left(\pi^{(3)}(e^{(3)})\right)\right) = e$, and set $e^{(4)} = \pi^{(4)}(e^{(3)})$. Remember that B is an arbitrary Borel subgroup of G.

LEMMA. *Each orbit of B in $G \cdot e^{(4)}$ contains $Z_i^{(4)}$ in its closure with the exception of the orbit $Bg \cdot e^{(4)}$, where $g \in G$ is chosen such that $B = gB_ig^{-1}$.*

PROOF: It is enough to prove the lemma for $B = B_i$. We have $e \in \mathbf{P}(N_i) \subset \mathbf{P}(M)$. Considering $S_i^{(4)}$ as a subset of $X^{(4)}$ via the inclusion $S_i^{(4)} = B_i \star^{B_i} S_i^{(4)} \hookrightarrow X_i^{(4)} \subset X^{(4)}$, one sees that $e^{(4)} \in S_i^{(4)}$. Let T_i be a maximal torus of B_i, and choose $w \in N_G(T_i) \setminus T_i$, where $N_G(T_i)$ is the normalizer of T_i in G. The set of fixed points of T_i in $\mathbf{P}(N_i)$ has two connected components: ${}^{B_i}\mathbf{P}(M)$ and the fixed point e_i of T_i in $Z_i \setminus \{d_i\}$. One can easily check that $\overline{T_i \cdot y} \ni e_i$ for all $y \in B_i \cdot e$, and therefore $\overline{T_i \cdot z} \ni d_i^{(4)}$ for all $z \in B_i \cdot e^{(4)}$. The closure of $B_i \cdot e^{(4)}$ in $X_i^{(4)}$ is $S_i^{(4)}$, thus it does not contain $Z_i^{(4)}$. From the Bruhat decomposition $G = B_iwB_i \cup B_i$, the orbits of B_i in $\overline{G \cdot e^{(4)}} \setminus \overline{B_i \cdot e^{(4)}}$ are of the form $B_iwb \cdot e^{(4)}$, with $b \in B_i$. We have $\overline{T_iwb \cdot e^{(4)}} = \overline{wT_ib \cdot e^{(4)}} \ni w \cdot d_i^{(4)}$, which implies that $\overline{B_iwb \cdot e^{(4)}} \supset \overline{B_iw \cdot d_i^{(4)}} = Z_i^{(4)}$. ∎

§3. Why the examples constructed in §1 are not schemes

Let X be one of the Artin-Moišezon varieties with a G-action constructed in §1. X contains a certain orbit $Z = \pi^{-1}(Z^{(5)})$ isomorphic to $\mathbf{P}_1$. Denote by X^0 the open orbit of G in X.

LEMMA. *The closure of each orbit of B in X^0 contains Z.*

PROOF: Denote by $(X^{(4)})^0$ and $(X^{(5)})^0$ the open orbits of G in $X^{(4)}$ and $X^{(5)}$. From the lemma of §2, since $r \geq 2$, the closure of each B-orbit in $(X^{(4)})^0$ contains at least one of the $Z_i^{(4)}$ $(i = 1, \ldots, r)$. The $Z_1^{(4)}, \ldots, Z_r^{(4)}$ all project down to $Z^{(5)}$ in $X^{(5)} = X^{(4)}/\Gamma$. Thus every orbit of B in $(X^{(5)})^0$ contains $Z^{(5)}$ in its closure. The lemma follows taking into consideration how X and Z are constructed from $X^{(5)}$ and $Z^{(5)}$. ∎

This lemma together with the following proposition shows that X is not an algebraic variety (algebraic variety = scheme).

PROPOSITION. *Let G be a semisimple (complex connected) group and B be a Borel subgroup of G. Let X be a normal algebraic G-variety. Choose two orbits Y and Z such that the dimension of Z is strictly positive. Then there exists at least one orbit of B in Y whose closure does not contain Z.*

PROOF: First we show it suffices to consider the case where Z is complete. By a result of Sumihiro [**Su**], one can embed X as an open G-stable subvariety of a normal algebraic complete G-variety $\overline{X}$. From a result of Bialynicki-Birula [**BB**], the closure of Z in $\overline{X}$ contains a complete orbit of strictly positive dimension. We can replace Z by this orbit.

The isotropy groups G_z ($z \in Z$) are then parabolic subgroups of G. Choose $z \in Z$ and a parabolic subgroup P of G opposed to G_z which contains B. Set $L = G_z \cap P$ and denote P^u the unipotent radical of P. From [**BLV**], there exists a (locally closed) affine subvariety W of X with the following properties:

- W contains z and is stable by L;
- the action of G on X induces an open immersion $P^u \times W \hookrightarrow X$.

The open subvariety P^uW of X is stable by P, and therefore also by B. Each orbit of B in the complement to P^uW does not contain Z in its closure. We are left to show that Y is not contained in P^uW.

Choose $y \in Y$. Denote by $q: G \to Y$ the morphism defined by $q(g) = g \cdot y$ ($g \in G$), and set $S = q^{-1}(W \cap Y)$. If Y were contained in $P^uW \xrightarrow{\sim} P^u \times W$, then the action of G on Y would induce an isomorphism $P^u \times (W \cap Y) \xrightarrow{\sim} Y$ and the multiplication in G an isomorphism $P^u \times S \xrightarrow{\sim} G$. Since G is affine, we would have that $P^u \backslash G \simeq S$ is affine, which is not the case (a homogeneous space under a reductive group is affine if and only if its isotropy groups are reductive). ∎

REMARK. The quotient of a projective variety by a finite group is a projective variety. Thus $X^{(4)}$ is not a projective variety.

§4. Closing comments and questions

We have found the examples of §1 (and other similar examples) in a natural way as a consequence of a systematic study of the almost homogeneous normal algebraic varieties of $SL_2(\mathbf{C})$ ([**LV**], [$\mathbf{MJ}_1$], [$\mathbf{MJ}_2$]). Since it seems to us that these examples are of interest in themselves, we wanted to make this article independent of the general studies. (However by choosing this concise formulation we hide intuition that comes from the larger perspective.)

Also we would like to mention that the construction of §1 is related to the procedure used by M. Nagata to construct the first known example of a non-quasiprojective algebraic variety ([**Nag**]). For other works dealing with three-dimensional almost homogeneous $SL_2(\mathbf{C})$-varieties see [**MU**] and [**Nak**].

All the examples of §1 have an infinite number of orbits. This suggests the following question: Does there exist a smooth compact connected Artin-Moišezon variety which is not an algebraic variety and which has an action of a reductive group with a finite number of orbits? Another question suggested by the construction given here: Does there exist a smooth compact connected Artin-Moišezon variety which is not an algebraic variety and on which a reductive group acts with an open orbit such that the isotropy is connected?

References

[**A**] M. Artin, *Algebraization of formal moduli: II. Existence of modifications*, Ann. of Math. **91** (1970), 88–135

[**BB**] A. Bialynicki-Birula, *On the action of* SL_2 *on complete algebraic varieties*, Pacific J. Math. **86** (1980), 53–58

[**BLV**] M. Brion, D. Luna, Th. Vust, *Espaces homogènes sphériques*, Invent. Math. **84** (1986), 617–632

[**C**] H. Cartan, *Quotient d'un espace analytique par un groupe d'automorphismes*. In: *Algebraic geometry and topology: a symposium in honor of S. Lefschetz*, Princeton Univ. Press, 1957

[**CA**] W.L. Chow, K. Kodaira, *On analytic surfaces with two independent meromorphic functions*, Proc. N.A.S. USA **38** (1952), 319–325

[**GA**] P. Griffiths, J. Adams, *Topics in Algebraic and Analytic Geometry*, Princeton Univ. Press, 1974

[Ha] R. Hartshorne, *Algebraic Geometry*, Springer-Verlag, 1977

[Hi] F. Hirzebruch, *Some problems on differentiable and complex manifolds*, Ann. of Math. **60** (1954), 213–236

[LV] D. Luna, Th. Vust, *Plongements d'espaces homogènes*, Comment. Math. Helv. **58** (1983), 186–245

[M] B.G. Moišezon, *On n-dimensional compact varieties with n algebraically independent meromorphic functions I, II, III*, Amer. Math. Soc. Translations, ser. 2, **63** (1967)

[MJ$_1$] L. Moser-Jauslin, *Normal* SL_2/Γ*-embeddings*, Thesis, Univ. of Geneva (1987)

[MJ$_2$] L. Moser-Jauslin, *Smooth Embeddings of* SL_2 *and* PGL_2, to appear

[M-S] S. Müller-Stach, *Bimeromorphe Geometrie von dreidimensionalen Moišezon Mannigfaltigkeiten*, Diplomarbeit, Universität Bayreuth (1987)

[MU] S. Mukai, H. Umemura, *Minimal rational threefolds.* In: *Algebraic Geometry, Lecture Notes in Math.* **1016**, Springer-Verlag, 1983

[Nag] M. Nagata, *On the embedding problem of abstract varieties in projective varieties*, Mem. Coll. Sci. Univ. Kyoto, Ser A, Math. **30** (1956), 71–82

[Nak] T. Nakano, *On equivariant completions of three-dimensional homogeneous spaces of* $SL_2(\mathbf{C})$, to appear in Jap. J. of Math

[Si] C.L. Siegel, *Meromorphe Funktionen auf kompakten analytischen Mannigfaltigkeiten*, Nach. Akad. Wiss. **4**, Göttingen (1955), 71–77

[Su] S. Sumihiro, *Equivariant Completion*, J. Math. Kyoto Univ. **14** (1974), 1–28

Domingo Luna
Institut Fourier
Université de Grenoble
F-38402 Saint-Martin-d'Hères
FRANCE

Lucy Moser-Jauslin et Thierry Vust
Section de Mathématiques
Université de Genève
2-4, rue du Lièvre, Case postale 240
CH-1211 Genève 24
SWITZERLAND

ON THE TOPOLOGY OF CURVES IN COMPLEX SURFACES

WALTER D. NEUMANN

Introduction

We describe the use of the link at infinity as a tool to classify complex affine plane curves. This is an exposition of results of [**N2**]. We give an overview of these results in §1, describe them precisely in §2, and give more examples in §4. §3 describes one of the topological ingredients: a substitute for the "Milnor fibration at infinity" when the latter does not exist.

The basic philosophy is that a three-manifold has a quite transparent structure given by its "toral decomposition," and that for a three-manifold link of a complex analytic object this structure is induced from the structure in the complex object, and thus tells one a lot about it.

The same philosophy is useful also in other situations, and in §5 we make some remarks on links of divisors in surfaces, which we apply to show how the link at infinity of an affine homology plane determines the compactification divisor. This part of the paper is independent from the foregoing; it expands on a short digression in the conference talk.

§1 Affine plane curves

Let V be a reduced algebraic curve in $\mathbf{C}^2$. Then the intersection of V with any sufficiently large sphere S^3 about the origin in $\mathbf{C}^2$ is transverse, and gives a well-defined link (S^3, L), called the *link at infinity of* $V \subset \mathbf{C}^2$. We denote it $\mathcal{L}(V, \infty)$. It turns out to be a very useful tool for studying the topology of plane curves. Some of our results need V to be "regular," or at least "regular at infinity," so we must begin by saying what this means.

Let $f\colon \mathbf{C}^2 \to \mathbf{C}$ be a polynomial map. A fiber $f^{-1}(c)$ is called *regular* if nearby fibers "look like" $f^{-1}(c)$ in the sense that, for some neighborhood D of $c \in \mathbf{C}$, $f \mid f^{-1}(D)\colon f^{-1}(D) \to D$ is a locally trivial C^∞ fibration. The fiber $f^{-1}(c)$ is *regular at infinity* if nearby fibers "look like it at infinity;" that is, for some neighborhood D of c and some compact subset K of $\mathbf{C}^2$, $f \mid f^{-1}(D) - K$ is a locally trivial fibration. "Regular" is equivalent to "regular at infinity and non-singular." Only finitely many fibers of f are irregular, and the fibers regular at infinity all define the same link at infinity up to isotopy, which we call the *regular link at infinity* of f and denote by $\mathcal{L}(f, \infty)$. If all fibers of f are regular at infinity we say f is *good*.

Proposition 1.1. *Any one of the following conditions implies f is good.*

(a) *Some fiber of f is reduced and has a knot as its link at infinity—i.e., it is connected at infinity* ([**N-R**, Lemma 7.1], [**N2**, Remark 2.3]).
(b) *$\mathcal{L}(f, \infty)$ is a fiberable link* ([**N-R**, Theorem 6.1], [**N2**, Remark 2.3]).
(c) *$\|x\|\,\|\text{grad}\, f(x)\|$ is bounded away from zero outside some compact subset of $\mathbf{C}^2$ (*MALGRANGE*, unpublished, quoted in* [**P**]*; see* [**B**] *for a weaker version).*

Condition (b) is in fact equivalent to goodness, since a good polynomial has a "Milnor fibration at infinity."

On the other hand, non-good polynomials are common. The easiest example is $f(x, y) = xy^2 + y$, a polynomial with no singularities. The fiber $f^{-1}(0)$ is not regular at infinity since it has three components at infinity while nearby fibers only have two. $\mathcal{K}(f, \infty)$ is the 2-component link consisting of an unknot together with a $(2, -1)$ cable on it (so both components are unknotted and they have mutual linking 2). As predicted by (b) above, this is not a fiberable link.

The application of links at infinity to algebraic curves is based on the following results.

Theorem 1.2. *If $V \subset \mathbf{C}^2$ is a regular curve (i.e., a regular fiber of its defining polynomial) then its link at infinity $\mathcal{L}(V, \infty)$ determines $V \subset \mathbf{C}^2$ as an embedded C^∞-manifold up to smooth isotopy.*

Conjecturally, V non-singular suffices for this theorem. The next results give some evidence for this.

First note that the degree $\deg(V)$ of V is only well defined relative to a fixed affine structure on $\mathbf{C}^2$; changing the embedding $V \subset \mathbf{C}^2$ by a non-affine automorphism of $\mathbf{C}^2$ generally changes the degree. (E.g., the curve $x^2 + y^3 = 1$ of degree 3 is taken by the algebraic automorphism $(x, y) \mapsto (x + y^2, y)$ of $\mathbf{C}^2$ to the curve $(x + y^2)^2 + y^3 = 1$ of degree 4.) Differently expressed, deg(V) depends on a choice of compactification $\mathbf{C}^2 \subset \mathbf{P}^2 = \mathbf{C}^2 \cup \mathbf{P}^1$: it is the algebraic intersection number with $\mathbf{P}^1$ of the closure $\overline{V}$ of V in $\mathbf{P}^2$. Denote by $\mathcal{DEG}(V)$ the set of all possible values of deg(V) and $d(V) := \min \mathcal{DEG}(V)$.

Theorem 1.3. *Suppose $V \subset \mathbf{C}^2$ is a reduced curve (so $\mathcal{L}(V, \infty)$ is defined), regular or not. Then*

(a) *$\mathcal{L}(V, \infty)$ determines $\mathcal{DEG}(V)$, and hence $d(V)$.*

(b) *$\mathcal{L}(V, \infty)$ determines $\chi(V) - \sum_p \mu_p$, where χ denotes the euler characteristic and the sum is the sum of the Milnor numbers at the singularities of V.*

The above results would be of little use if we could not tell which links actually occur as links at infinity. But we do in fact have strong conditions on a link for it to occur, and the invariants involved in Theorem 2 are then very easy to compute. We shall describe this in the next section, but first we give an example application.

Suppose we want to classify all non-singular algebraic $V \subset \mathbf{C}^2$ of given genus with one point at infinity (this case—one point at infinity—can also be dealt with using ABHYANKAR's approach [**A**,**A-S**]). The set of candidates for the link $\mathcal{L}$ at infinity for V is very easy to compute. Namely, the theory tells us that $\mathcal{L}$ is a knot $\mathcal{L} = \mathcal{O}\{p_1, q_1; p_2, q_2; \dots; p_k, q_k\}$ (this means the (p_k, q_k)-cable on the (p_{k-1}, q_{k-1})-cable on the ... on the (p_1, q_1)-cable on the unknot $\mathcal{O}$), with cabling coefficients $p_i > 1$, $q_i > 0$ satisfying $\gcd(p_i, q_i) = 1$, $1 < q_1 < p_1$, and $q_i < p_i p_{i-1} q_{i-1}$ for $i > 1$. If this $\mathcal{L}$ is the link at infinity of V then the precise version of Theorem 1.3 (see Theorem 2.3) says that (possibly after an algebraic automorphism of $\mathbf{C}^2$), V is given by a polynomial of degree $P = p_1 \cdots p_k$ and has genus

$$g = \frac{1}{2}\left(\sum_1^k (q_i(p_i - 1)p_{i+1} \cdots p_k) - P + 1\right) \quad .$$

For each g this gives finitely many $\mathcal{L}$ and, for each $\mathcal{L}$, since we know in what degree we must look, it is then a finite calculation to work out the polynomials that give this $\mathcal{L}$. We are only interested in the polynomial up to algebraic automorphisms of $\mathbf{C}^2$, which considerably simplifies calculations. For low g we get the following list:

$\mathcal{L}$	d	g	f
$\mathcal{O}$	1	0	x
$\mathcal{O}\{3,2\}$	3	1	x^3+y^2+ax
$\mathcal{O}\{5,2\}$	5	2	$x^5+y^2+ax^3+bx^2+cx$
$\mathcal{O}\{4,3\}$	4	3	$x^4+y^3+y(ax^2+bx+c)+dx^2+ex$
$\mathcal{O}\{3,2;2,3\}$	6	3	$f_1(x,y)$
$\mathcal{O}\{7,2\}$	7	3	$x^7+y^2+ax^5+bx^4+cx^3+dx^2+ex$
$\mathcal{O}\{5,3\}$	5	4	$x^5+y^3+y(ax^3+bx^2+cx+d)+ex^3+fx^2+gx$
$\mathcal{O}\{3,2;2,5\}$	6	4	$f_2(x,y)$
$\mathcal{O}\{9,2\}$	9	4	$x^9+y^2+ax^7+bx^6+cx^5+dx^4+ex^3+fx^2+gx$
$\mathcal{O}\{3,2;3,2\}$	9	4	$f_3(x,y)$

where

$$f_1(x,y)=(x^3+y^2)^2+ax^4+bx^3+\frac{1}{4}a^2x^2+cx+axy^2+by^2+dy,\qquad (d\neq 0),$$

$$f_2(x,y)=(x^3+y^2)^2+ax^4+bx^3+cx^2+dx+axy^2+by^2+exy+fy,\qquad (e\neq 0),$$

$$\begin{aligned}f_3(x,y)=(x^3+y^2)^3&+3ax^7+bx^6+3a^2x^5+2abx^4+cx^3+a^2bx^2+dx\\&+6ax^4y^2+2bx^3y^2+3a^2x^2y^2+2abxy^2+(c-a^3)y^2-6axy^4+by^4,\\&\qquad (d\neq a^4-8ac).\end{aligned}$$

In each case V is given by $f(x,y)=h$, where h is an additional constant, suitably chosen to make V non-singular. If any one of the parameters $a,b,\dots,h$ is non-zero, then it can be made to equal 1 by a linear transformation, so the true number of parameters in each case is one less than appears.

We can also use the results to classify singular curves with given topology. For example, if V has one point at infinity, topological genus 0, and has one normal crossing, then a nearby fiber will be smooth of genus 1. Thus, by the second line of the table, V is given by $x^3+y^2+ax=b$, for some a and b. The singular points are $(\pm\sqrt{-a/3},0)$. One of them must lie on V, so $b=\pm\sqrt{-4a^3/27}$. If $a=0$ then the singularity is a cusp, so $a\neq 0$. A simple change of coordinates now puts the equation in the form $x^3+y^2+x^2=0$.

For curves with more than one point at infinity there are infinitely many $\mathcal{L}$ for each genus, so similar classifications become harder. For example, we have not yet classified the smooth curves of genus 0 with 2 points at infinity, although is seems feasible, and the classification of the regular ones is easy (we give an irregular one in Example 4.1 below): up to algebraic automorphisms the polynomials $f(x, y)$ whose regular fibers are annuli are

$$\begin{aligned} f(x,y) &= x^p y^q \ , \quad \text{or} \\ f(x,y) &= (yx^r + a_{r-1}x^{r-1} + \cdots + a_0)^q x^{p-rq} \end{aligned}$$

with $0 < q \leq p$, $0 < rq < p$, $a_0 \neq 0$ and $\gcd(p, q) = 1$.

§ 2 Toral links and splice diagrams

A link at infinity is always a *toral link*, that is, it can be built up by iterated cabling operations from the unknot. In **[E-N]** such links are classified by *splice diagrams* and we will need to give a quick review of these. The splice diagram for a link at infinity is just a coding of the Newton- Puiseux pairs at infinity; the results could therefore be formulated without reference to the link at infinity at all, but we will not do this.

We first fix terminology about cabling. Given a component K of a link $\mathcal{L} = (S^3, L)$, let $N(K)$ denote any closed solid torus neighborhood of K which is disjoint from all other components of L. An (α, β) *cable* on K is a simple closed curve $K(\alpha, \beta)$ which lies on some $\partial N(K)$ and is homologous in $N(K)$ to αK and has linking number β with K (so α and β must be coprime). Several (α, β) cables on K are *parallel* if they lie on some common $\partial N(K)$ and are mutually disjoint. An (α, β) *cabling operation on the component* K *of* $\mathcal{L}$ is the operation of either replacing K by, or adding, some number $d \geq 1$ of parallel (α, β) cables on K, to obtain a new link.

Splice diagrams are certain decorated trees used to represent toral links. The "decorations" consist of integer weights at the ends of some edges. Also, some of the vertices are drawn as arrowheads; they correspond to components of the link. When constructing a toral link by iterated cabling, we will start from the "n-component Hopf link" $\mathcal{H}_n = (S^3, K_1 \cup \ldots \cup K_n)$ consisting of n fibers of the Hopf fibration. This is the link at infinity for $x^n - y^n = 0$. We represent this link by the splice diagram with n arrowheads:

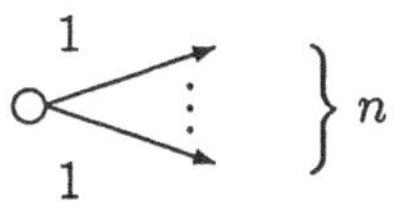

.

If $\boxed{\Gamma}\longrightarrow$ is a splice diagram for a link $\mathcal{L}$ and the indicated arrowhead corresponds to a component K, then each of the diagrams

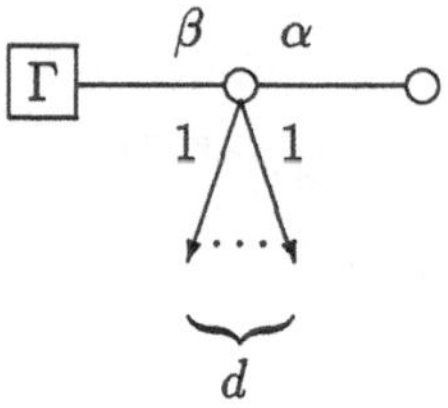

and

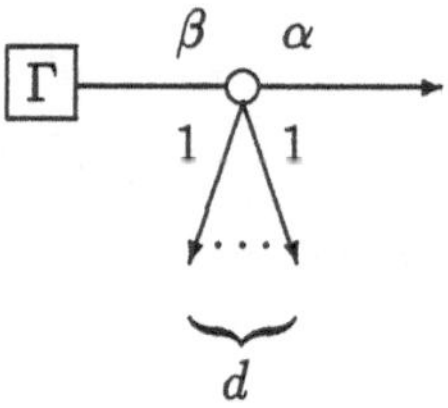

represents the result of doing an (α, β) cabling operation to $\mathcal{L}$; namely, either replacing component K by, or adding, d parallel (α, β) cables on K. One can thus construct a splice diagram for any toral link. In [**E-N**] it is shown that the splice diagram determines the link. The splice diagram itself is determined by the link only up to certain operations. For instance, an edge with weight 1 at one end and a leaf at its other end can be discarded (a *leaf* is a non-arrowhead vertex of valency 1—*valency* is the number of incident edges at the vertex). Vertices of valency 2 can be ignored. Also certain changes of sign of the weights in a splice diagram are allowed.

If we have constructed a toral link by iterated cabling from $\mathcal{H}_n$ then we call the vertex of its splice diagram which comes from this $\mathcal{H}_n$ the *root vertex* of the splice diagram and speak of a *rooted splice diagram*. There may be many ways of picking a root vertex in a splice diagram for a toral link. They represent different ways of constructing the link by iterated cabling.

For example, the following rooted diagrams give two different constructions for one link; the root vertex is marked as a "•" in each:

(2.1)

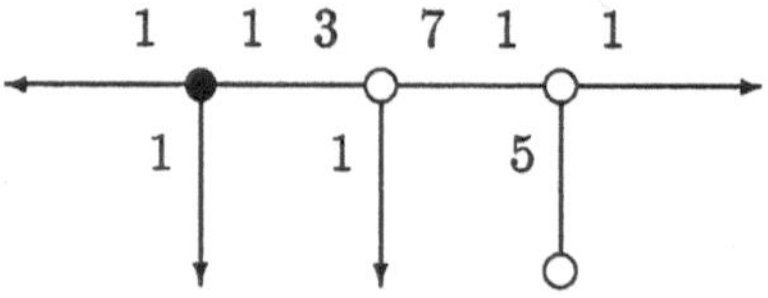

,

(2.2)

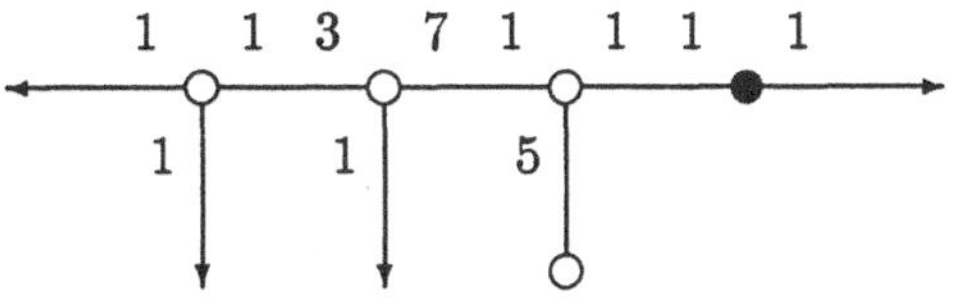

In the first case we start from $\mathcal{H}_3 = (S^3, K_1 \cup K_2 \cup K_3)$ and add a $(7,3)$-cable $K_3(7,3)$ and then replace K_3 by its $(5,1)$-cable $K_3(5,1)$. In the second case we start from $\mathcal{H}_2$, add the $(5,1)$-cable $L = K_2(5,1)$, and then successively add $L(3,7)$ and $L(1,1)$ (note that $L(1,1)$ is drawn on the boundary of a "thinner" solid torus than $L(3,7)$, since the solid torus must be disjoint from the already existing $L(3,7)$. Yet another splice diagram for the same link is:

(2.3)

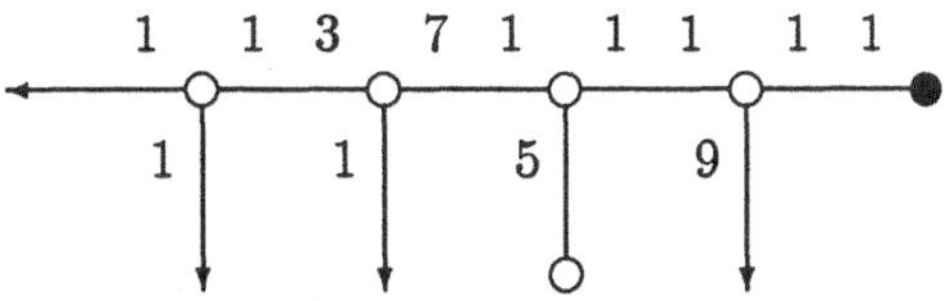

This involves a redundant cabling operation—the $(9,1)$-cabling is a complicated way of creating an unknot—we say this splice diagram is "not minimal."

To indicate the correspondence between the splice diagram and Newton-Puiseux pairs at infinity, we discuss this example from the point of view of links at infinity. The above link is the link at infinity for

$$x(x+y)(x^3+y^7)(x+y^5) = c\,.$$

By a (non-linear) change of coordinates in $\mathbf{C}^2$ we can write this as

$$(x-y^5)(x-y^5+y)((x-y^5)^3+y^7)x = c\,.$$

Another change of coordinates gives the equation

$$(x-(y+x^9)^5)(x-(y+x^9)^5+(y+x^9))((x-(y+x^9)^5)^3+(y+x^9)^7)x = c\,.$$

There is a correspondence between these equations and the diagrams (2.1)–(2.3) which should be clear, and which is an instance of a general result. To describe this result we need more terminology.

In a rooted splice diagram we call a weight on an edge *near* or *far* according as it is an the end of the edge nearest to or farthest from the root vertex. By construction, at most one near weight at a vertex is unequal to 1.

A rooted splice diagram Γ is an *RPI splice diagram* if

(a) all near weights are positive and

(b) for any edge with weights α_0 and α_1 on it and weights $\alpha_2, \ldots, \alpha_m$ adjacent to but not on it,

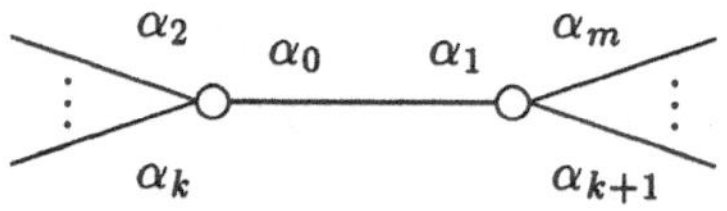

,

one has $\alpha_0\alpha_1 - \alpha_2 \ldots \alpha_m < 0$, that is, the *edge determinant* is negative.

(RPI stands for *reverse Puiseux inequalities*. The classical *Puiseux inequalities*, which are necessary and sufficient conditions for a link to be a link of a plane curve singularity, are equivalent to the link having a splice diagram satisfying (a) together with the reverse of (b)—equivalently all weights and all edge determinants are positive.) A link $\mathcal{L} = (S^3, L)$ which can be represented by an RPI splice diagram Γ will be called an *RPI link*.

A (rooted) splice diagram Γ for a link $\mathcal{L} = (S^3, L)$ determines certain curves in $S^3 - L$, called *virtual components* for $\mathcal{L}$, which can be described as follows. Γ describes an iterative cabling construction for $\mathcal{L}$. Redo this construction to create extra components (which are never themselves cabled on) as follows: start from $\mathcal{H}_{n+1} = (S^3, K_0 \cup K_1 \cup \cdots \cup K_n)$ instead of $\mathcal{H}_n$ (so K_0 is an extra component) and, whenever a cabling operation is performed, include an extra parallel cable (which is never cabled on later). The extra components created this way are *virtual components of the first kind*. Any K which was replaced in some cabling operation during the construction is a *virtual component of the second kind*. These virtual components clearly correspond naturally with the non-arrowhead vertices in the splice diagram Γ. In particular, K_0 is the virtual component for the root vertex. We shall denote the virtual component corresponding to a vertex v by S_v and denote $l_v = \text{link}(S_v, L)$: the linking number of this virtual component with the link.

In [**E-N**] a quick way of computing such linking numbers is described:

Lemma 2.1 ([**E-N**, Sect. 10]). *The linking number of any two compo-*

nents (virtual or genuine) for the link is the product of all weights adjacent to, but not on, the simple path in the splice diagram connecting the corresponding vertices.

For example, in the diagram (2.1) above, the virtual component S_{root} for the root vertex has linking numbers 1, 1, 7, 5, with each of the four components of the link, so $l_{root} = 1 + 1 + 7 + 5 = 14$. Similarly, for the diagram (2.2) $l_{root} = 1 + 15 + 5 + 5 = 26$ and for (2.3) $l_{root} = 1 + 135 + 45 + 45 = 226$.

An RPI splice diagram for which the l_v are all non-negative will be called a *regular splice diagram* (it suffices to require $l_v \neq 0$ just for vertices v which are adjacent to arrowheads). It turns out that if some RPI diagram for a link $\mathcal{L}$ is regular then they all are; we then call $\mathcal{L}$ a *regular toral link*.

An RPI link $\mathcal{L}$ has many different RPI diagrams describing it, but there are finitely many *minimal* RPI diagrams, where minimality means that there are no "obviously redundant" cablings involved in the corresponding iterative construction of $\mathcal{L}$. For a link at infinity these minimal RPI diagrams differ only in the placement of the root vertex (this is a special property, not valid for arbitrary RPI links).

Theorem 2.2. *Given a reduced algebraic curve $V \subset \mathbf{C}^2$, any embedding $\mathbf{C}^2 \subset \mathbf{P}^2$ (equivalently, an affine structure on $\mathbf{C}^2$) determines an RPI splice diagram Γ for $\mathcal{L}(V, \infty)$. Every minimal RPI splice diagram Γ for $\mathcal{L}(V, \infty)$ is realized by some embedding $\mathbf{C}^2 \subset \mathbf{P}^2$. If V is regular at infinity then Γ is a regular splice diagram, i.e., $\mathcal{L}(V, \infty)$ is a regular toral link (conjecturally the converse is true also).*

Now pick any embedding $\mathbf{C}^2 \subset \mathbf{P}^2 = \mathbf{C}^2 \cup \mathbf{P}^1$. Let Γ be the corresponding RPI splice diagram for $\mathcal{L}(V, \infty)$.

Theorem 2.3.

(a) *The number $n_\infty(V) := \overline{V} \cap \mathbf{P}^1$ of points at infinity of V is the valency n of the root vertex in Γ.*

(b) $\deg(V)$ *is the linking number l_{root} of the virtual component for the root vertex with the link.*

(c) $\chi(V) = \sum_v (2 - \delta_v) l_v + \sum_p \mu_p$, *where the first sum is over the non-arrowhead vertices of Γ, δ_v denotes the valency of vertex v, and the second sum is over the singularities p of V, and μ_p is the Milnor number at p.*

The proofs are given in detail in [**N2**]. Very briefly, the fact that a link $\mathcal{L}(V,\infty)$ is given by an RPI diagram follows easily from the Puiseux inequalities for the singularities "at infinity" of $\overline{V} \subset \mathbf{P}^2 = \mathbf{C}^2 \cup \mathbf{P}^1$. The fact that this RPI diagram is regular if V is regular then follows by comparing the values for $\chi(V)$ predicted by knot theory on the one hand and by elementary algebraic geometry on the other. The formulae of Theorem 2.3 come out along the way.

The algebraic realizability of any minimal RPI splice diagram for $\mathcal{L}(V,\infty)$ is a (reasonably canonical) computation with Newton diagrams, once one has shown that different minimal RPI diagrams for $\mathcal{L}(V,\infty)$ differ only in the position of the root vertex. The latter is a combinatorial argument using an additional property of links that arise as links at infinity. (If V has just one component at infinity, the realizability of the—in this case unique—minimal RPI diagram was proved in very different language by ABHYANKAR and SINGH [**A-S**]).

Actually, more is true. Define an RPI diagram Γ to be *reduced* if the RPI splice diagram $\overline{\Gamma}$ obtained by adding an extra arrow (with weight 1) at the root vertex is a minimal RPI splice diagram. For example, the diagram (2.3) is reduced but not minimal.

Theorem 2.4. *Given a reduced algebraic curve $V \subset \mathbf{C}^2$, the splice diagrams for $\mathcal{L}(V,\infty)$ corresponding to embeddings $\mathbf{C}^2 \subset \mathbf{P}^2$ are precisely the reduced RPI splice diagrams for $\mathcal{L}(V,\infty)$.*

The proof is essentially the same as the proof in [**N2**] for minimal RPI diagrams. Alternatively, it is not hard to show that one can realize the link determined by $\overline{\Gamma}$ as $\mathcal{L}(V \cup L,\infty)$, where L is the image of some algebraic embedding of $\mathbf{C}$ in $\mathbf{C}^2$. There is then an embedding $\mathbf{C}^2 \subset \mathbf{P}^2$ realizing $\overline{\Gamma}$, by the result for minimal RPI diagrams, and removing L realizes Γ.

In particular, in Theorem 1.3, $\mathcal{DEG}(V)$ is the set of values of l_{root} for all reduced RPI diagrams for $\mathcal{L}(V,\infty)$. In the example of diagram (2.1) this is the set of all multiples of 14, 27, 29, or of a number of the form $25n+1$ with $n>0$.

§ 3 Multilinks

We have said nothing yet about the proof of Theorem 1.2. If the defining polynomial f for V is good, then, as mentioned in Proposition 1.1, there is a "Milnor fibration at infinity." The intersection of V with any large enough ball is isotopic to a fiber of this Milnor fibration, and it is well known that a fiber of a fibered link is determined up to isotopy by the link, proving 1.2 in this case. The key to proving 1.2 in general is a replacement for the Milnor fibration at infinity when f is not good:

Theorem 3.1.

(a) *A regular plane curve V can be recovered up to isotopy from a suitable spanning surface $F \subset S^3$ for its link at infinity $\mathcal{L} = (S^3, L)$ by attaching a collar out to infinity in $\mathbf{C}^2$ to the boundary ∂F.*

(b) *F is the fiber of a fibered multilink $\mathcal{L}'$ which is uniquely determined by $\mathcal{L}$.*

A *multilink*, see **[E-N]**, is a link whose components have been assigned integer multiplicities; it is *fibered* if its *exterior*—the complement of an open tubular neighborhood of the link—is fibered over the circle by a map whose degree on any curve is the linking number of the curve with the link, taking multiplicities into account.

The uniqueness of the multilink $\mathcal{L}'$ associated with $\mathcal{L}$ is a purely topological fact: in **[N2]** it is shown that the boundary of the fiber of a fibered multilink (with no zero multiplicities) determines the multilink. The existence of $\mathcal{L}'$ is a topological consequence of the fact that $\mathcal{L}$ is a regular toral link, but a more direct analytic proof, due to L. RUDOLPH, is also given in **[N2]**. $\mathcal{L}'$ is a quite mysterious object from an algebraic point of view: even if one disregards its multiplicities it is itself usually not a link at infinity of an algebraic curve.

§ 4 Examples

Example 4.1. Let p, q, r be positive integers with p and r coprime and $r > 1$, $p < (q+1)r$. Consider $f(x,y) = (x^q y + 1)^r + x^p$. Then $V = f^{-1}(0)$ is smooth, and the splice diagram Ω for its link $\mathcal{L}$ at infinity is

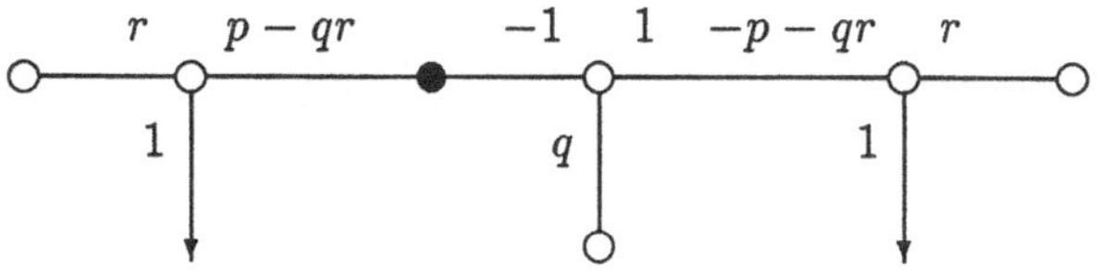

The values of l_v, reading across the diagram, are p, pr, $qr + r$, 0, $-pr$, $-p$ (in particular, $\mathcal{L}$ is not regular). Theorem 2.3 shows that the euler

characteristic of V is 0, and since V has no closed components, it must be either an annulus or the disjoint union of a punctured sphere and a punctured torus. The latter possibility is ruled out because the two components have non-trivial linking number. Thus V is an annulus. This f has another irregular fiber, $f^{-1}(1)$, whose link at infinity has splice diagram

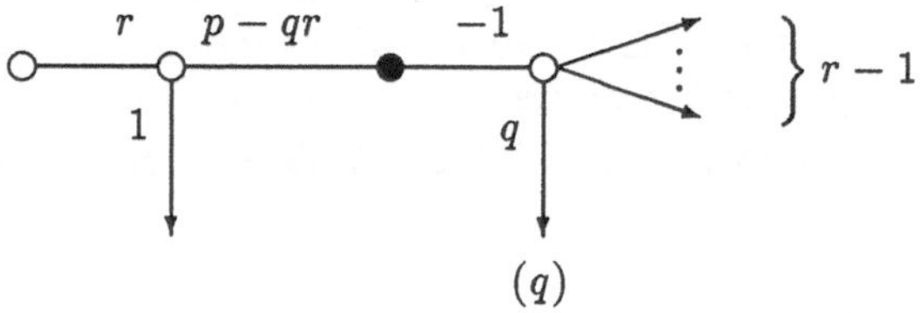

,

where the q in parentheses indicates that the corresponding link component has multiplicity q. The regular link at infinity for f has splice diagram

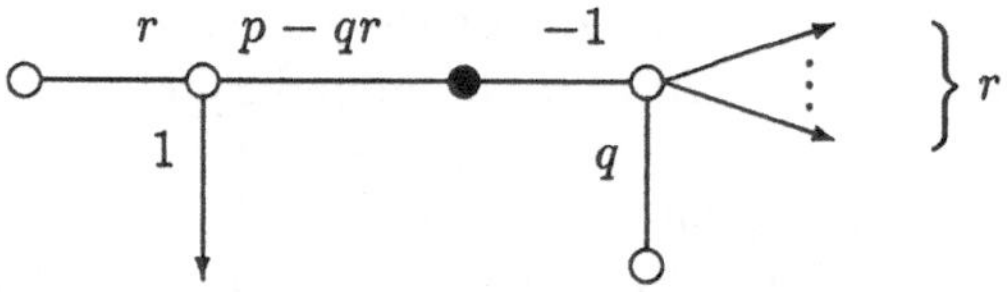

,

so Theorem 2.3 implies that a regular fiber of f has euler characteristic $p(1-r)$ and hence topological genus $(p-1)(r-1)/2$. Finally, the multilink associated with this polynomial has splice diagram

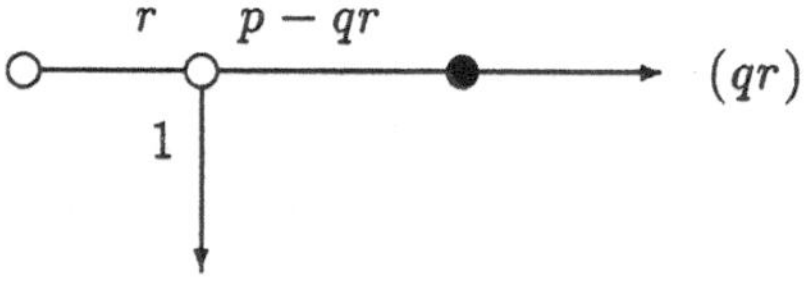

,

If $p < qr$ then the underlying link of this multilink is not realizable as a link at infinity, however one orients its components.

Example 4.2. Consider $f(x, y) = x^p y^q + y$ with $\gcd(p, q) = \gcd(p, q-1) = 1$, $p \geq 1$, $p \geq 2$. All fibers of f are smooth. The only irregular fiber is $f^{-1}(0)$. It is the disjoint union of an annulus and a disk; its link at infinity has splice diagram

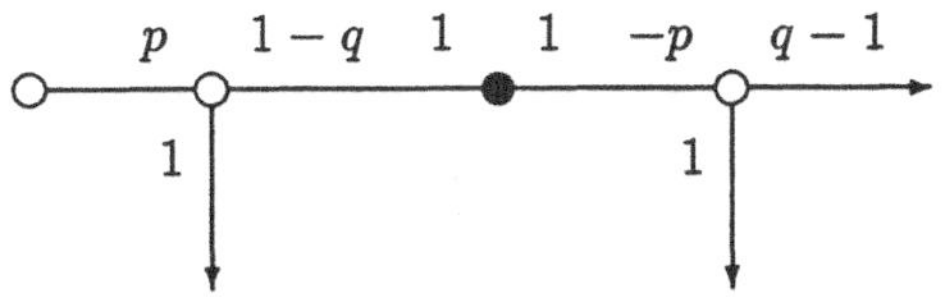

The regular fiber is a twice punctured surface of genus $(p-1)/2$. Its link at infinity has RPI splice diagram

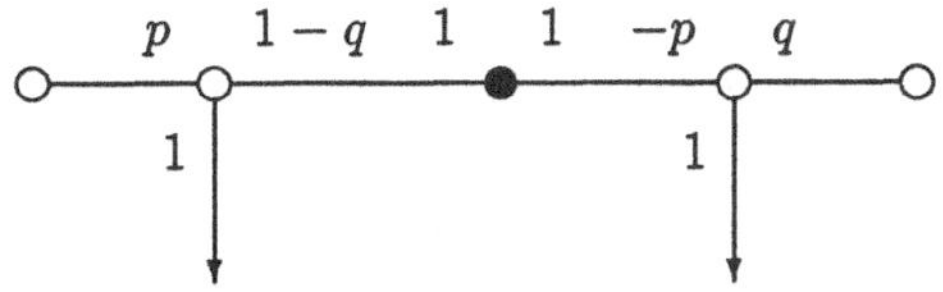

The multilink corresponding to $\mathcal{L}$ has diagram

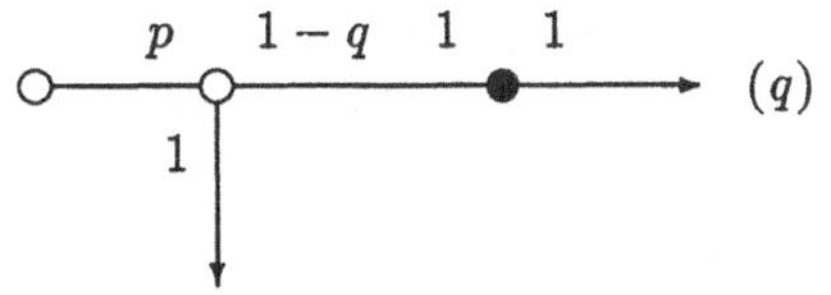

The underlying link of this multilink is not realizable as a link at infinity, however one orients its components, if $q > 2$.

§ 5 Links of compact curves and compactification divisors of homology planes

A reduced curve D in a nonsingular complex surface V is *good* if the only singularities of D are normal crossings. It is, in addition, *minimal* if no component can be blown down preserving the goodness property; that is, each nonsingular rational component C of D with selfintersection $C \cdot C = -1$ must intersect the rest of D in at least 3 points. Any reduced curve D in V can be made good by blowing up singularities and then be made minimal by blowing down curves which contravene the minimality condition.

Let N be a closed regular neighborhood of D in V. If D is good, the topology of the pair (N, D) can be coded in the usual way by the dual intersection graph Γ. This is the graph with a vertex v_C for each irreducible component C of D and an edge for each normal crossing in D (this edge connects the vertices which correspond to the components which cross at this point; these two vertices may be equal) and with two integer weights associated to each vertex v_C: the euler number e_C of the normal bundle $\nu_V(C)$ and the genus g_C of C. Note that $e_C = C \cdot C - 2m_C$, where m_C is the number of self-crossings of C, so $e_C = C \cdot C$ if C is nonsingular. The pair (N, D) can be reconstructed from this data by plumbing according to Γ (see e.g., [**N1**]).

Assume until further notice that D is a minimal good curve and is topologically connected. The smooth oriented 3-manifold $M = \partial N$ is called the *link of* D. In [**N1**] we showed that M determines Γ, and hence the topology of (N, D), in each of the following cases:

- D is the exceptional divisor of the resolution of a singularity.
- D is the fiber of a holomorphic map of V to a curve.

The same result holds in several other situations. One situation of interest is the minimal good compactification divisor of a normal affine surface X: such a surface can be embedded in a compact projective surface V in such a way that X is the complement $V - D$ of a minimal good curve D in V and no singularity of V lies on D; the question naturally arises as to whether X determines D.

The following theorem, implicit in [**N1**], is the underlying result. Recall that a *chain* in Γ is a connected subgraph all of whose vertices v_C have genus weight $g_C = 0$ (i.e., C is rational) and valency ≤ 2 in Γ (*valency* is the number of incident edges, with edges that begin and end at v_C counted twice). We say that Γ *has negative chains* if the euler weight e_C satisfies $e_C < 0$ for every vertex on a chain of Γ. By minimality of D, we then in fact have $e_C < -1$ on chains, except for the special case $\Gamma = -1$ 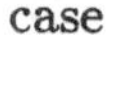.

Theorem 5.1. *If Γ is not -1*  *and has negative chains then the oriented 3-manifold M determines Γ (in fact, if M is not a lens space then its "oriented fundamental group" determines Γ). Moreover, M is a prime 3-manifold with respect to connected sum.*

A compact complex surface V which contains a nonsingular rational curve C with $C \cdot C \geq 0$ is either rational, or ruled, or a blow-up of such a surface ([**B-P-V**, Prop. V.4.3]). L. J. McEwan has showed me that in the special case $\Gamma = -1$, V will be rational by similar techniques. In any case, we see:

Corollary 5.2. *Theorem 5.1 always applies to show that M is prime and determines Γ if the ambient surface V is not rational, ruled, or a blow-up of such a surface.*

It is more interesting to have results that make no *a priori* assumptions about the ambient surface beyond compactness.

Theorem 5.3. *If V is compact and $\mathrm{H}_1(M;\mathbf{Z}) = 0$ and $M \neq S^3$, then Γ is a tree with negative chains, so Theorem 5.1 applies. In fact, all euler weights in Γ are negative except possibly for weights 0 at one node or at two adjacent nodes (a "node" is a vertex of valency ≥ 3).*

One interest of this result is that it applies to the minimal good compactification divisor of a complex affine homology plane (for more on complex affine homology planes see [**P-tD**]). However there are many curves D covered by Theorem 2 which cannot occur as such a compactification divisor. For example, the curve with graph

$\Gamma =$ 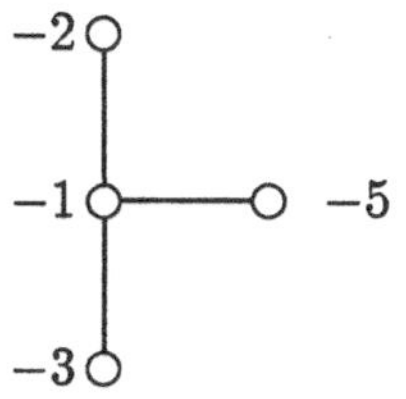

cannot be the compactification divisor of a complex affine homology disk (since if it were, then its link M would bound a smooth homology 4-disk; but M is the POINCARÉ sphere, which is well known not to do so: its ROCHLIN invariant is non-zero, for instance). On the other hand, this curve can occur in a compact surface. Take a cubic with one cusp in $\mathbf{P}^2$ and resolve the cusp. The resulting curve has graph

$\Gamma =$ 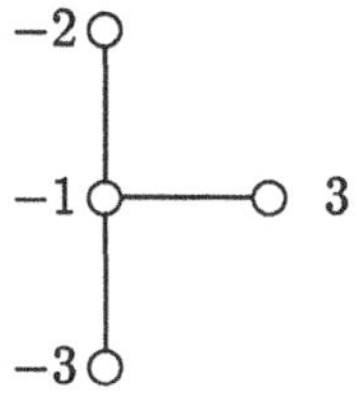

,

and now blowing up 8 times on the proper transform of the cubic reduces its selfintersection number from 3 to -5. Without the restriction on $\mathrm{H}_1(M)$, Theorem 5.3 is false. In fact, *any non-trivial connected sum of lens spaces occurs*, so M is certainly not always prime. Moreover, infinitely many Γ occur for each such M. If $M = \mathbf{P}^3\mathbf{R}$ then these Γ include:

$$-2\circ\;,\quad 2\circ\;,\quad 0\circ\!\!-\!\!\!-\!\!\overset{l}{\circ}\!\!-\!\!\!-\!\!\circ\;-2 \qquad \text{with } l \neq -1, \text{ etc.}$$

They are all linear, but the length can be arbitrary. They are easy to classify. In fact, all these Γ other than $-2\circ$ are mutually equivalent by complex analytic blowing up and blowing down, so up to blowing up and down there are only two cases. This behavior is general for $H_1(M)$ finite:

Theorem 5.4. *For any bound m there exist only a finite number of exceptional M with $|H_1(M)| \leq m$ for which Γ may fail to have negative chains. But for these exceptions there are still only finitely many Γ up to analytic blowing up and down.*

The proofs of 5.3 and 5.4 will appear in [**N3**].

Acknowledgements

This research was partially supported by the N.S.F., and some of the research and most of the writing was done at the Max-Planck-Institut für Mathematik in Bonn. Their support is gratefully acknowledged.

References

[**A**] S. S. Abhyankar, *On Expansion Techniques in Algebraic Geometry*, Tata Institute of Fundamental Research, Lectures on Mathematics and Physics **57** (1977).

[**A-S**] Shreeram S. Abhyankar and Balwant Singh, "Embeddings of certain curves," Amer. J. Math. **100** (1978), 99–175.

[**B-P-V**] W. Barth, C. Peters, and A. van de Ven, *Compact Complex Surfaces*, Ergebnisse der Mathematik und ihre Grenzgebiete (3) **4**, Springer (1984).

[**B**] S. A. Broughton, "Milnor numbers and the topology of polynomial hypersurfaces," Invent. Math. **92** (1988), 217–242.

[**E-N**] David Eisenbud and Walter Neumann, *Three-dimensional Link Theory and Invariants of Plane Curve Singularities*, Ann. of Math. Studies **101**, Princeton Univ. Press (1985).

[**M**] John Milnor, *Singular Points of Complex Hypersurfaces*, Ann. of Math. Studies **61**, Princeton Univ. Press (1968).

[N1] Walter D. Neumann, "A calculus for plumbing applied to the topology of complex surface singularities and degenerating complex curves," Trans. Amer. Math. Soc. **268** (1981), 299–344.

[N2] Walter D. Neumann, "Complex algebraic plane curves via their links at infinity," (to appear in Invent. Math.).

[N3] Walter D. Neumann, "Topology of curves in compact complex surfaces," in preparation.

[N-R] Walter D. Neumann and Lee Rudolph, "Unfoldings in knot theory," Math. Annalen **278** (1987), 409–439, and "Corrigendum: Unfoldings in knot theory," ibid. **282** (1988), 349–351.

[P-tD] T. Petrie and T. tom Dieck, "Homology planes. An announcement and survey," these Proceedings.

[P] F. Pham, "La descente des cols par les ongletes de Lefschetz avec vues sur Gauss-Manin," *Systèmes Différentiels et Singularités*, Astérisque **130** (1985), 11–47.

[S] Masakazu Suzuki, "Propriétés topologiques des polynômes de deux variables complexes, et automorphismes algébriques de l'espace $\mathbf{C}^2$," J. Math. Soc. Japan **26** (1974), 241–257.

[Z-L] M. G. Zaidenberg and V. Y. Lin, "An irreducible, simply connected curve in $\mathbf{C}^2$ is equivalent to a quasihomogeneous curve," (English translation), Soviet. Math. Dokl. **28** (1983), 200–204.

[W] F. Waldhausen, "On irreducible 3-manifolds that are sufficiently large," Ann. of Math. **87** (1968), 56–88.

Walter D. Neumann
Department of Mathematics
Ohio State University
Columbus, OH 43210
USA

THE TOPOLOGY OF ALGEBRAIC QUOTIENTS

GERALD W. SCHWARZ

§ 0. Introduction

(0.0) This is an expanded version of my talk of the above title at the Rutgers Conference on Topological Methods in Algebraic Transformation Groups, April 4–8, 1988. I report on recent work which shows there is a strong interconnection between the theory of quotients in the algebraic category (reductive groups) and in the topological category (compact groups). It turns out that one can represent quotient spaces of reductive groups as quotient spaces of compact groups. The precise method of doing this has important topological consequences for quotients of reductive groups and also provides some new insight into quotients of compact groups. I hope that my exposition will make these results more accessible to all.

(0.1) Let G be a reductive complex algebraic group (see §2) and S an affine complex algebraic G-variety. Then one can define a natural "quotient mapping" $\pi_{S,G}: S \to S/\!\!/G$, where S is the "quotient space" of S by G (see §3). Kempf and Ness **[KN]** showed that there is a real algebraic subset M_S of S such that $S/\!\!/G$ is homeomorphic to M_S/K, where K is a maximal compact subgroup of G (§4). Moreover, Neeman **[Ne]** showed that there is a K-equivariant retraction of S onto M_S (§5, §7). Combined with work of Oliver **[Ol]** (see also **[KPR]**) one obtains strong topological consequences. For example, if S is contractible, so is $S/\!\!/G$. In §6 we sketch some results of Procesi-Schwarz **[PS]** which show how to use the above results to gain new insight into quotients by compact groups.

(0.2) Many of the techniques and results above extend to the case of representations of real reductive groups, as has been shown recently by R. Richardson and P. Slodowy **[Sl]**.

Research partially supported by the NSF and NSA.

(0.3) I thank E. Bierstone for valuable discussions concerning §7 and P. Slodowy for pointing out an error in an earlier version of (4.7).

§ 1. Quotients by Compact Groups

(1.0) We discuss properties of quotients of representations and real algebraic varieties by compact Lie groups.

(1.1) Let K be a compact Lie group and $\rho: K \to \mathrm{GL}(W)$ a real representation of K. By a theorem of Hilbert and Hurwitz ([**Weyl**, Ch.VIII, §14]) the (graded) algebra $\mathbf{R}[W]^K$ of K-invariant polynomials on W is finitely generated, say by homogeneous elements $p_1, \dots, p_d$. Let

$$p = (p_1, \dots, p_d): W \to \mathbf{R}^d,$$

let I be the ideal of relations of the p_i in $\mathbf{R}[y_1, \dots, y_d]$, and let Z be the corresponding algebraic subset of $\mathbf{R}^d$. Then $X := \operatorname{Im} p$ is a semialgebraic subset of Z ([**BiMl**, §1]). Let W/K denote the orbit space (quotient topology), and let $\overline{p}: W/K \to X$ denote the mapping induced by p.

(1.2) Proposition (see [**Schw**], [**Br**, Ch.1, §3]).

(1) *The maps p and $\overline{p}$ are proper, hence X is closed in $Z \subseteq \mathbf{R}^d$.*

(2) *The map p separates disjoint closed K-orbits, and $\overline{p}: W/K \to X$ is a homeomorphism.*

(1.3) Remark ([**Schw**]): It can be shown that the map $\overline{p}$ is even a *diffeomorphism*, if one defines $C^\infty(W/K) := C^\infty(W)^K$ and $C^\infty(X) := C^\infty(\mathbf{R}^d)|_X$.

(1.4) As one can see from the above, the theory of quotients of representations by compact groups is quite nice. There is one mystery, however. How does one describe X? The set Z is determined by $\mathbf{R}[W]^K$ (in fact, viewed abstractly, Z is just a realization of the real maximal spectrum of $\mathbf{R}[W]^K$). The description of X involves *inequalities*. For example, if $K = \{\pm 1\}$ acts via multiplication on $\mathbf{R}$, then $(X \subseteq Z) \simeq (\mathbf{R}^+ \subseteq \mathbf{R})$. How does one find the inequalities? We return to indicate the solution to this problem in §6.

(1.5) Remark: Let S be a real affine algebraic K-variety. Then there is an equivariant closed algebraic embedding $S \hookrightarrow W$ where W is a representation space of K (see [**Kr**, II.2.4]). The image $S' \subseteq W$ can be defined by a single invariant equation (take sums of squares!), hence $S/K \simeq S'/K \simeq X' := p(S') = X \cap Z'$, where Z' is an algebraic subset of Z. Thus the inequalities determining $X \subseteq Z$ determine $X' \subseteq Z'$.

§ 2. Reductive Complex Algebraic Groups

(2.0) In order to keep matters as topological as possible, we define reductive groups via Weyl's unitarian trick. The references [**Weyl**], [**Chev**], and [**Helg**] contain all the results needed in this section.

(2.1) Let K be a compact Lie group. Then K has a faithful embedding $K \hookrightarrow \mathrm{GL}_n(\mathbf{R})$ for some n (Peter-Weyl theorem). Identify K with its image. Then K is a real algebraic subgroup of $\mathrm{GL}_n(\mathbf{R})$. Let I_K denote the ideal defining K, and let $G := K_{\mathbf{C}}$ denote the complex zeroes of I_K in $\mathrm{GL}_n(\mathbf{C})$.

(2.2) Proposition. *Let K, G etc. be as above. Then*

(1) *G is a complex algebraic group, which, up to isomorphism, is independent of the embedding of K chosen.*
(2) *$\mathfrak{g} = \mathfrak{k} + i\mathfrak{k}$, where $\mathfrak{g} = \mathrm{Lie}(G)$ = Lie algebra of G; $\mathfrak{k} = \mathrm{Lie}(K)$.*
(3) *K is dense in G (Zariski topology of G).*
(4) *$G = K \cdot P \simeq K \times P$ where $P = \exp(i\mathfrak{k}) \subseteq \mathrm{GL}_n(\mathbf{C})$, and P is diffeomorphic (via log) to $i\mathfrak{k}$.*
(5) *K is a maximal compact subgroup of G.*

(2.3) Examples: (1) $K = \mathrm{O}_n(\mathbf{R})$, $G = \mathrm{O}_n(\mathbf{C})$. Then I_K is defined by the equation: $AA^t = I$.

(2) $K = \mathrm{U}_n(\mathbf{C}) \subseteq \mathrm{GL}_{2n}(\mathbf{R})$, $G = \mathrm{GL}_n(\mathbf{C}) \subseteq \mathrm{GL}_{2n}(\mathbf{C})$. Here

$$K \simeq \left\{ \begin{pmatrix} A & B \\ -B & A \end{pmatrix} \in \mathrm{GL}_{2n}(\mathbf{R}) : A^tA + B^tB = I; A^tB - B^tA = 0 \right\}.$$

Changing from co-ordinates $x_1, \ldots, x_n, y_1, \ldots, y_n$ on $\mathbf{R}^{2n}$ to the co-ordinates $z_j = x_j + iy_j$ and $z'_j = x_j - iy_j$ on $\mathbf{C}^{2n}$, the image $K \hookrightarrow \mathrm{GL}_{2n}(\mathbf{C})$ becomes all matrices

$$\begin{pmatrix} g & 0 \\ 0 & (g^{-1})^t \end{pmatrix} \qquad , g \in K.$$

Since $\dim_{\mathbf{R}} K = n^2 = \dim_{\mathbf{C}} \mathrm{GL}_n(\mathbf{C})$,

$$K_{\mathbf{C}} = \left\{ \begin{pmatrix} g & 0 \\ 0 & (g^{-1})^t \end{pmatrix} \right\} \subseteq \mathrm{GL}_{2n}(\mathbf{C}),$$

where $g \in \mathrm{GL}_n(\mathbf{C})$ is arbitrary.

(2.4) REMARK ([**Hoch**]): If G is any real Lie group with finitely many components, then G has maximal compact subgroups, and all are G-conjugate. Moreover, if K is a maximal compact subgroup, then G/K is contractible.

(2.5) DEFINITION (Non-Standard): A complex algebraic group G is *reductive* if $G \simeq K_{\mathbf{C}}$ for some compact Lie group K.

(2.6) REMARK: Since algebraic groups have finitely many components, it follows that G is reductive if and only if $\mathfrak{g} = \mathfrak{k} \oplus i\mathfrak{k}$ where $\mathfrak{k}$ is the Lie algebra of a maximal compact subgroup. In particular, G is reductive if and only if its identity component G^0 is reductive.

(2.7) EXAMPLE: The additive group $(\mathbf{C}, +)$ is not reductive since its maximal compact subgroup ($= \{0\}$!) has dimension zero.

(2.8) PROPOSITION (Weyl's Unitary Trick). *Let $G = K_{\mathbf{C}}$ be reductive, and let $\rho: G \to \mathrm{GL}(V)$ be a representation.*

(1) *V is a completely reducible G-module (i.e., G is linearly reductive).*
(2) *The algebra $\mathbf{C}[V]^G$ is finitely generated.*

PROOF: Since K is Zariski dense in G, a subspace V' of V is K-stable if and only if it is G-stable. Since K is linearly reductive, so is G. The polynomial invariants of a representation of a linearly reductive group are finitely generated ([**Kr**, II.3.2] or [**Weyl**, Ch.VIII, §14]), hence $\mathbf{C}[V]^G = \mathbf{C}[V]^K$ is finitely generated.

§ 3. Quotients by Reductive Groups

(3.0) One of the difficulties of dealing with actions of reductive complex algebraic groups is the presence of non-closed orbits. The orbit space is then highly non-Hausdorff. The usual way of proceeding is to define the quotient to be the space of closed orbits. One then studies the quotient space and the fibers of the canonical mapping to it.

(3.1) Let $\rho: G \to \mathrm{GL}(V)$ be a representation of the reductive complex algebraic group G. Let $V/\!\!/G$ denote the complex affine variety corresponding to $\mathbf{C}[V]^G$, and let $\pi_{V,G}: V \to V/\!\!/G$ be the morphism dual to the inclusion $\mathbf{C}[V]^G \hookrightarrow \mathbf{C}[V]$. We can realize $V/\!\!/G$ in an analogous manner to the orbit space W/K of §1: Let $q_1, \ldots, q_e$ be homogeneous generators of $\mathbf{C}[V]^G$, let $q = (q_1, \ldots, q_e): V \to \mathbf{C}^e$, and let $Y \subseteq \mathbf{C}^e$ denote the variety defined by the relations of the q_i (see (1.1)). Then $Y \simeq V/\!\!/G$.

(3.2) PROPOSITION. *Let $q: V \to Y \subseteq \mathbf{C}^e$ be as above. Then*

(1) $\operatorname{Im} q = Y$.
(2) *q separates disjoint closed G-invariant algebraic subsets of V.*
(3) *If $v \in V$, then $\overline{Gv} \setminus Gv$ is a union of orbits of dimension less than $\dim Gv$.*
(4) *Every orbit contains a unique closed orbit in its closure.*
(5) *The map q sets up a bijection between closed orbits of V and points of Y (and similarly for $\pi_{V,G}$, V and $V/\!\!/G$).*

PROOF (see [**MumF**, Ch.1, §2], [**Lu1**], [**Kr**, II.3.2]): Property (3) holds for any algebraic group. Clearly, (4) and (5) are implied by (1), (2) and (3).

Note that, in the complex case, the "orbit space" is determined by $\mathbf{C}[V]^G$. There is no problem analogous to (1.4).

(3.3) EXAMPLE: Let $G = \mathrm{GL}_n(\mathbf{C})$ act by conjugation on $V = \mathrm{M}_n(\mathbf{C})$. Let $\mathrm{tr}: V \to \mathbf{C}$ denote the trace function. It is classical that $\mathbf{C}[V]^G$ is generated by the functions

$$q_j: \mathrm{M}_n(\mathbf{C}) \ni A \mapsto \mathrm{tr}(A^j) \in \mathbf{C}, \quad j = 1, \ldots, n.$$

Let $q = (q_1, \ldots, q_n)$. The *zero fiber* or *null cone* is $q^{-1}(0)$, and consists of all nilpotent matrices. The only closed orbit in the zero fiber is that of the 0 matrix.

(3.4) REMARKS: (1) Let S be an affine G-variety. Then, as in (1.5), we have an equivariant closed embedding $\iota: S \hookrightarrow V$ for some representation space V of G. Moreover, $\iota^*\mathbf{C}[V]^G = \mathbf{C}[S]^G$ ([**Kr**, II.3.2]). It follows that $\mathbf{C}[S]^G$ is finitely generated, and we have a commutative diagram:

$$\begin{array}{ccc} S & \xrightarrow{\ \iota\ } & V \\ \big\downarrow{\scriptstyle \pi_{S,G}} & & \big\downarrow{\scriptstyle \pi_{V,G}} \\ S/\!\!/G & \xrightarrow{\ \bar\iota\ } & V/\!\!/G \end{array}$$

where $\bar\iota$ is a closed embedding and $\pi_{S,G}$ and $S/\!\!/G$ are defined as in (3.1).

(2) More generally, let $f: S \to S'$ be an equivariant morphism of affine G-varieties. Then f^* maps $\mathbf{C}[S']^G$ to $\mathbf{C}[S]^G$ and induces a morphism $f/\!\!/G: S/\!\!/G \to S'/\!\!/G$. There is a commutative diagram analogous to the one above.

(3.5) The quotient space $S/\!\!/G$ we have constructed is the reasonable one in the category of algebraic varieties, and $\pi_{S,G}$ has the following universal property: "If $\varphi: S \to Z$ is a morphism of varieties, constant on G-orbits, then φ is a composition $\phi \circ \pi_{S,G}$, where $\phi: S/\!/G \to Z$ is a morphism." One can ask if $\pi_{S,G}$ is universal in other categories:

(3.6) THEOREM (Luna [**Lu2**], [**Lu3**]). *$\pi_{S,G}: S \to S/\!/G$ is universal in the category of Hausdorff (even T_1) spaces and in the category of complex analytic varieties.*

We give a proof of the above results of Luna in the next section.

§ 4. Quotients via Compact Groups

(4.0) We recount in this section results of Kempf and Ness [**KN**] which show how to represent quotients $S/\!/G$, considered as topological spaces, as quotients M_S/K, where $G = K_{\mathbf{C}}$ and $M_S \subseteq S$. Neeman [**Ne**] showed that M_S is a K-equivariant deformation retract of S. We exploit this result in §5.

(4.1) Let $\rho: G \to V$ be a representation of the reductive complex algebraic group G. Let K be a maximal compact subgroup of G, and let $\langle\ ,\ \rangle$ be a K-invariant hermitian form on V with associated norm $\|\ \|$. Let $v \in V$ and consider the function $F_v: G \to \mathbf{R}^+$ sending g to $\|gv\|^2$.

(4.2) THEOREM. *Let G, V, v, etc. be as above.*

(1) *F_v has a critical point if and only if Gv is closed.*
(2) *All critical points of F_v are minima.*

Assume that $F_v(e)$ is a minimum. Then

(3) $Kv = \{v' \in Gv : \|v'\| = \|v\|\}$.
(4) $G_v = (K_v)_{\mathbf{C}}$.

(4.3) REMARKS: Parts (1), (2) and (3) are due to Kempf and Ness. Part (4) was observed by Dadok and Kac [**DaK**].

Before going into the details of the proof of the theorem, we draw some important consequences.

(4.4) COROLLARY (Matsushima's Theorem). *Suppose that H is an algebraic subgroup of G such that G/H* (see [**Humph**, IV]) *is an affine variety. Then H is reductive.*

PROOF: As in §2, we may find an equivariant closed embedding $G/H \hookrightarrow V$ for some representation space V of G. The image of G/H is a closed orbit Gv. By (4.2.4), there is a point gv whose isotropy group is reductive. Hence H is reductive. ∎

(4.5) Let $M = \{v \in V : (dF_v)_e = 0\}$. Then $M = \{v \in V : T_vGv \perp v\}$, and we have the following equivalent conditions for membership in M:

(4.5.1) $\langle Av, v\rangle = 0$ for all $A \in \mathfrak{g}$,

(4.5.2) $\langle Av, v\rangle = 0$ for all $A \in \mathfrak{k}$,

where we consider $\mathfrak{k} \subseteq \mathfrak{g} \subseteq \mathrm{End}(V)$. We call M the *Kempf-Ness set* of V. Note that M is a real algebraic cone in V.

(4.6) DEFINITION: Let S be an affine G-variety (which we may assume is equivariantly embedded in a representation V of G). We define the Kempf-Ness set M_S of S to be $M \cap S$. Of course, M_S, considered as a subset of S, depends upon the embedding and norm on V.

(4.7) COROLLARY. *The composition $M_S \hookrightarrow S \rightarrow S/\!\!/G$ is proper and induces a homeomorphism $M_S/K \rightarrow S/\!\!/G$.*

PROOF (cf. [**Sl**]): Suppose that $\pi_{S,G}|_{M_S}$ is proper. Then the induced mapping $\pi'\colon M_S/K \rightarrow S/\!\!/G$ is also proper, and it is 1-1 and onto by (3.2), (3.4) and (4.2). Since M_S/K and $S/\!\!/G$ are locally compact Hausdorff, it follows that π' is a homeomorphism.

Clearly, $\pi_{S,G}|_{M_S}$ is proper if $\pi_{V,G}|_M$ is proper. If $\pi_{V,G}|_M$ is not proper, then there is a sequence v_j in $M \setminus \{0\}$ such that $\lim_{j\to\infty} \|v_j\| = \infty$ while $\{\pi_{V,G}(v_j)\}$ remains in a compact subset of $V/\!\!/G$. Set $m_j := v_j/\|v_j\|$. Then $\pi_{V,G}(m_j) \rightarrow \pi_{V,G}(0)$ as $j \rightarrow \infty$, and, since $\|m_j\| = 1$ for all j, we may assume that $m_j \rightarrow m_0 \in M$ as $j \rightarrow \infty$. Then m_0 lies in the null cone $\pi_{V,G}^{-1}(\pi_{V,G}(0))$, which implies that $m_0 = 0$, a contradiction. ∎

(4.8) COROLLARY (Luna [**Lu2**], [**Lu3**]).

(1) *Let F be a closed G-invariant subset of S (classical topology). Then $\pi_{S,G}(F)$ is closed in $S/\!\!/G$.*

(2) (Theorem (3.6)) $\pi_{S,G}$ is the universal quotient in the category of Hausdorff spaces and in the category of complex analytic varieties.

PROOF: Part (1) is easy, since $\pi_{S,G}(F) \simeq (F \cap M_S)/K$ is closed in $S/\!\!/G \simeq M_S/K$. Thus a subset of $S/\!\!/G$ is closed if and only if its inverse image in S is closed, hence the standard classical topology on $S/\!\!/G$ is the same as the quotient topology. Let $\varphi\colon S \rightarrow Z$ be continuous and G-invariant, where Z is Hausdorff. If $s, s' \in S$ and $s' \in \overline{Gs}$, then $\varphi(s') \in \overline{\{\varphi(s)\}}$, hence $\varphi(s) = \varphi(s')$ by Hausdorffness of Z. Thus φ is constant on the fibers of $\pi_{S,G}$, and $\varphi = \phi \circ \pi_{S,G}$, where $\phi\colon S/\!\!/G \rightarrow Z$ is continuous since $S/\!\!/G$ has the quotient topology.

It remains to establish that $\pi_{S,G}$ is universal for the category of complex analytic varieties. We assume that $S \hookrightarrow V$ is an equivariant closed linear embedding, and we assume a bit of familiarity with complex analysis on the part of the reader (see [**GR**] and [**Wh**]). Let $\mathcal{H}(\bullet)$ denote the holomorphic functions on $\bullet$. Let $q\colon V \to Y \subseteq \mathbf{C}^e$ be as in (3.2), let U be open in $q(S) \subseteq Y$, and let $f \in \mathcal{H}(U')^G$ where $U' = q^{-1}(U) \cap S$. We need to show that $f = q^*(h)$ where $h \in \mathcal{H}(U)$. There is a unique h on U such that $f = q^*h$, and h is continuous by our results above.

Since analyticity is a local property, we may assume that $U = D \cap q(S)$, where D is the polydisk $\{y \in \mathbf{C}^e : |y_i - c_i| < \epsilon,\ i = 1, \ldots, e\}$ for some $c_1, \ldots, c_e \in \mathbf{C}$ and $\epsilon > 0$. Set $P = q^{-1}D$. Then P is polynomially convex, hence Stein, so there is an analytic function $\tilde{f}$ on P such that the restriction of $\tilde{f}$ to $P \cap S = U'$ is f. By averaging over K, we can arrange that $\tilde{f} \in \mathcal{H}(P)^K = \mathcal{H}(P)^G$. If $\tilde{f} = q^*\tilde{h}$ where $\tilde{h} \in \mathcal{H}(D \cap Y)$, then $h = \tilde{h}|_U$ is analytic. We may thus reduce to the case that $S = V$, $U = D \cap Y$ and $f \in \mathcal{H}(P)^G$. We present two proofs that h is holomorphic:

First proof: Let $L \subseteq U$ be compact, and let L' denote $q^{-1}(L) \cap M$. Since P is polynomially convex, there is a sequence of polynomials p_j converging to f uniformly on L'. Averaging over K, we may assume that the p_j are invariant: $p_j = q^*r_j$ where $r_j \in \mathbf{C}[y_1, \ldots, y_e]$. The r_j converge uniformly to h on L. Since $\mathcal{H}(U)$ is closed under uniform convergence on compact sets, we obtain that $h \in \mathcal{H}(U)$.

Second Proof: Let $U^0 = \{q(v) \in U : \operatorname{rank}(dq)_v = \dim Y\}$. Then U^0 is open in U, and its complement is a proper analytic subvariety. By the implicit function theorem, h is holomorphic on U^0. Since h is continuous on U and U has only normal singularities ([**Kr**, II.3.3]), h is actually holomorphic on U. ∎

(4.9) We now turn to the proof of Theorem (4.2): Let $\mathfrak{k}_v$ denote the Lie algebra of K_v. Recall that $\mathfrak{g} = \mathfrak{k} \oplus i\mathfrak{k}$. Let $Y \in i\mathfrak{k}$, and consider the function $a(s) := \|\exp(sY)v\|^2;\ s \in \mathbf{R}$.

(4.10) LEMMA. *Either $Yv = 0$ (in which case $a'' \equiv 0$), or $a''(s) > 0$ for all s.*

PROOF (see [**DaK**]): Since $Y \in i\mathfrak{k}$, Y is hermitian symmetric with respect to $\langle\ ,\ \rangle$, and we may choose an orthonormal basis $e_1, \ldots, e_n$ of V such that $Ye_j = m_j e_j$; $j = 1, \ldots, n$, where the m_j are real. If $v = \sum v_i e_i$, then

$$a(s) = \sum e^{2sm_j} |v_j|^2, \qquad \text{and}$$
$$a''(s) = \sum 4m_j^2 e^{2sm_j} |v_j|^2.$$

The only way for $a''(s)$ to vanish is for v_j to be zero whenever $m_j \neq 0$, i.e. for Yv to be zero. ∎

PROOF OF (4.2.2–4.2.4): The function F_v has a critical point at h if and only if F_{hv} has a critical point at the identity $e \in G$, so we may assume that F_v has a critical point at e. Let $g \in G$. By (2.2.4), g can be uniquely written as $k \exp(Y)$ where $k \in K, Y \in i\mathfrak{k}$. Now $\|gv\| = \|\exp(Y)v\| = a(1)$, where a is as above. Since $a'(0) = 0$ (by our hypothesis on F_v) and $a'' \geq 0$, we can only have $\|gv\| \geq \|v\|$. Thus F_v has a minimum at e, establishing (4.2.2). If $\|gv\| = \|v\|$, then $a'' \equiv 0$, which implies that $Yv = 0$. Hence $\exp(Y) \in G_v$ and $gv = kv$, establishing (4.2.3). If $gv = v$, then $k \in K_v$ and $Y \in i\mathfrak{k}_v$, so that $G_v = K_v \exp(i\mathfrak{k}_v) = (K_v)_{\mathbf{C}}$, establishing (4.2.4).

PROOF OF (4.2.1): Suppose that Gv is not closed. Then the Hilbert-Mumford theorem (see **[MumF]** or **[Kr]**) shows that there is a 1-parameter subgroup of G (i.e., a homomorphism $\lambda: \mathbf{C}^* \to G$) such that

$$\lim_{z \to 0} \lambda(z)v = v_0$$

exists and lies outside of Gv. In other words, if Gv is not closed, then there is already a copy of $\mathbf{C}^*$ in G such that $\overline{\mathbf{C}^* v} \not\subseteq Gv$. We need slightly more:

(4.11) LEMMA. *There is a* $\lambda: \mathbf{C}^* \to G$ *such that* $v_0 = \lim_{z \to 0} \lambda(z)v \notin Gv$ *and* $\lambda(\mathbf{S}^1) \subseteq K$.

($\mathbf{S}^1$ is, of course, the maximal compact subgroup of $\mathbf{C}^*$.) Let λ be as in (4.11), and let $Y \in i\mathfrak{k}$ denote the image of $1 \in \mathrm{Lie}(\mathbf{C}^*) = \mathbf{C}$ under $d\lambda$. Then, as in the proof of (4.10),

$$\|\lambda(e^s)v\|^2 = \|\exp(sY)v\|^2 = \sum_{j=1}^{n} e^{2m_j s}|v_j|^2.$$

Since $\lambda(e^s)v$ has a limit as $s \to -\infty$, we must have that $m_j|v_j| \geq 0$ for all j, and we have strict inequality for some j since the limit lies outside of $\lambda(\mathbf{C}^*)v$. Thus $\frac{d}{ds}\|\lambda(e^s)v\|^2 \neq 0$ when $s = 0$, and F_v is not critical at $e \in G$. ∎

PROOF OF (4.11): There is a parabolic subgroup $P(\lambda)$ canonically associated to λ (see **[MumF]** or **[Kr]**), where

$$P(\lambda) = \{g \in G : \lim_{z \to 0} \lambda(z)g\lambda(z^{-1}) \text{ exists}\}.$$

(For example, suppose that $G = \mathrm{GL}_n(\mathbf{C})$ and $\lambda(z) = \mathrm{diag}(z^{a_1}, \ldots, z^{a_n})$ where $a_1 \geq a_2 \geq \ldots \geq a_n$. Then $g = (g_{ij}) \in P(\lambda)$ if and only if $g_{ij} = 0$ whenever $i > j$ and $a_i \neq a_j$.) If $p \in P(\lambda)$, one easily sees that $\lim_{z \to 0} p\lambda(z)p^{-1}v = v_p$ exists and $v_p \notin Gv$. Note that $\lambda(\mathbf{C}^*) \subseteq P(\lambda)$.

It is a general fact that the intersection of two parabolics has maximal rank ([**Humph**, §28]). Now $P(\lambda)$ and $\overline{P(\lambda)}$ are parabolic, so that $P(\lambda) \cap \overline{P(\lambda)}$ has maximal rank. Moreover, it is clearly defined over **R**, so it is the complexification of a subgroup L of K (of maximal rank). By conjugacy of maximal tori ([**Humph**, §21]), there is a $p \in P(\lambda)$ such that $p\lambda(\mathbf{C}^*)p^{-1} \subseteq T_{\mathbf{C}}$, where T is a maximal torus of L. Then $p\lambda p^{-1} = \lambda'$ is easily seen to have the required properties. ∎

§ 5. Topology of Algebraic Quotients

(5.0) We begin with Neeman's deformation theorem. It allows us to apply topological results of Oliver concerning compact group actions to reductive group actions.

(5.1) Theorem. *Let $\rho: G \to \mathrm{GL}(V)$, K, M, etc. be as in §4. Then*

(1) *There is a K-equivariant deformation retraction φ_t of V to M, $0 \le t \le 1$, where $\varphi_0 = id$, $\varphi_1: V \to M$.*
(2) *The deformation retraction respects the fibers of $\pi_{V,G}$; in fact, $\{\varphi_t(v) : 0 \le t < 1\} \subseteq Gv$ and $\varphi_1(v) \in \overline{Gv}$.*

(5.2) Remark: Let S be a closed G-subvariety of V. Then (5.1.2) shows that the deformation retraction $V \to M$ restricts to a deformation retraction $S \to M_S$.

(5.3) We defer the proof of (5.1) to §7, but we say a few words now to present the ideas involved: Let $f_v = (dF_v)_e: \mathfrak{g} \to \mathbf{R}$. We may consider f_v as an element of the real dual $\mathfrak{g}^*$ of $\mathfrak{g}$. Since K is norm preserving, f_v vanishes on $\mathfrak{k}$, so we consider f_v as an element of $i\mathfrak{k}^* \simeq \mathfrak{k}^* \simeq \mathfrak{k}$, where the last identification is obtained from a K-invariant inner product $\{\ ,\ \}$ on $\mathfrak{k}$. Let $h(v) = \|f_v\|^2$, where the norm comes from $\{\ ,\ \}$. Note that $h(v) \ge 0$ and that $h(v) = 0$ if and only if $v \in M$.

Let ψ_t be the local 1-parameter group generated by $-\operatorname{grad} h$. (The gradient is taken relative to $\langle\ ,\ \rangle$.) In §7 we show that ψ_t exists for all $t \ge 0$, and that $\psi_t: \mathbf{R}^+ \times V \to V$ extends to a continuous mapping of $\mathbf{R}^+ \cup \{\infty\} \times V \to V$, where $\mathbf{R}^+ \cup \{\infty\} \simeq [0,1]$. Note that ψ_t is K-equivariant by construction. Moreover, one easily computes that $(dh)_v(w) = 0$ if $w \perp i\mathfrak{k}(v)$; in other words, $(\operatorname{grad} h)(v)$ points along $i\mathfrak{k}(v)$, so that the flow of ψ_t is along G-orbits.

As an immediate consequence of (5.1) and (5.2) we have

(5.4) Corollary. *Let S, G, M_S, etc. be as in (5.2). Then the inclusion $M_S \hookrightarrow S$ is a homotopy equivalence in the K-equivariant category. From the homeomorphism of $S/\!\!/G$ and M_S/K we obtain an induced homotopy equivalence*

$$S/\!\!/G \simeq S/K.$$

We can now apply results of Oliver:

(5.5) Theorem (Oliver [**Ol**]). *Let K be a compact Lie group and X, Y paracompact K-spaces of finite cohomological dimension with finitely many orbit types.*

(1) *Let $f: X \longrightarrow Y$ be a K-equivariant map which induces an isomorphism in cohomology. Then the quotient mapping $f/K: X/K \longrightarrow Y/K$ induces an isomorphism in cohomology.*
(2) *If X is acyclic (resp. contractible), then so is X/K.*

(5.6) Remark: All our K-spaces S, etc. satisfy the hypotheses of (5.5).

(5.7) Corollary (cf. [**KPR**]). *Let S, S' be affine G-varieties.*

(1) *Let $f: S \longrightarrow S'$ be a G-equivariant morphism which induces an isomorphism in cohomology. Then $f/\!\!/G: S/\!\!/G \longrightarrow S'/\!\!/G$ induces an isomorphism in cohomology.*
(2) *If S is acyclic (resp. contractible), then so is $S/\!\!/G$.*

§ 6. Inequalities Defining Orbit Spaces

(6.0) In §§4–5 we exploited the fact that quotients by complex reductive groups can also be obtained as quotients by their maximal compact groups. In order to solve the problem posed in (1.4) concerning quotients by compact groups, we reverse gears and complexify!

(6.1) Let $\rho: K \longrightarrow \mathrm{GL}(W)$, and $p = (p_1, \dots, p_d): W \longrightarrow X \subseteq Z \subseteq \mathbf{R}^d$ be as in §1. We want to find the inequalities defining X. By choosing an appropriate basis $x_1, \dots, x_n$ of W we may identify W with $\mathbf{R}^n$ so that K preserves the standard inner product (,). Let $A(w) = (\partial p_i/\partial x_j(w))$, $w \in W$. Then

$$(AA^t)_{ij} = (\operatorname{grad} p_i, \operatorname{grad} p_j) = \sum \frac{\partial p_i}{\partial x_k}\frac{\partial p_j}{\partial x_k} \in \mathbf{R}[W]^K \simeq \mathbf{R}[Z].$$

Hence there is a symmetric matrix valued function Grad on Z such that $\mathrm{Grad}(p(w)) = (AA^t)(w)$, $w \in W$.

(6.2) THEOREM (**[PS]**). $X = \{z \in Z : \text{Grad}(z) \geq 0\}$.

By $\text{Grad}(z) \geq 0$ we mean that the matrix is positive semidefinite. The discussion in (6.1) shows that $\text{Grad}(x) \geq 0$ for all $x \in X$. Note that the condition $B \geq 0$ for a symmetric matrix B is equivalent to a simultaneous set of inequalities $\{B_\alpha \geq 0 : \alpha \in \Gamma\}$ where B_α runs over the determinants of symmetric minors of B.

(6.3) REMARK: Let $\tilde{Z}$ denote the real maximal spectrum of $\mathbf{R}[W]^K$, and let $\tilde{X}$ denote the image of W in $\tilde{Z}$. We may think of Grad as a symmetric matrix $\widetilde{\text{Grad}}$ of elements of $\mathbf{R}[W]^K$. Then the points of $\tilde{Z}$ where $\widetilde{\text{Grad}}$ is positive semidefinite are exactly $\tilde{X}$. We may use one set of generators of $\mathbf{R}[W]^K$ to determine a matrix $\widetilde{\text{Grad}}$ and another set to realize $\tilde{Z}$ and $\tilde{X}$ as (semi-)algebraic subsets of some R^d. See (6.5) below for an application.

(6.4) EXAMPLE: Let $K = S_n$, the symmetric group on n-letters, act standardly on $W = \mathbf{R}^n$. Then $\mathbf{R}[W]^K = \mathbf{R}[\tau_1, \ldots, \tau_n]$ where $\tau_j(x) = \tau_j(x_1, \ldots, x_n) = \sum x_i^j$, $j \geq 0$. Since $\tau_1, \ldots, \tau_n$ are algebraically independent, $Z = \mathbf{R}^n$. Let A denote the Jacobian of $\tau = (\tau_1, \ldots, \tau_n)$: $W \to \mathbf{R}^n$. Then AA^t has ij-entry $ij\tau_{i+j-2}$, and AA^t is positive semidefinite if and only if the "Bezoutiant" matrix $\text{Bez} = (\text{Bez})_{ij} = (\tau_{i+j-2})$ is positive semidefinite. Thus

$$X = \tau(W) = \{y \in \mathbf{R}^n : \text{Bez}(y) \geq 0\}.$$

(6.5) EXAMPLE: Let $f(x) = x^n - b_1 x^{n-1} + \ldots + (-1)^n b_n$ be a real monic polynomial with roots $a_1, \ldots, a_n$. When are all the roots real? We actually just answered the question! We know that $b_j = \sigma_j(a)$, where σ_j is the jth elementary symmetric function and $a = (a_1, \ldots, a_n)$. Since the σ_j generate the algebra $\mathbf{R}[W]^K$ of example (6.4), we may apply remark (6.3):

(6.6) THEOREM (Sylvester). *The roots of f are all real if and only if* $\text{Bez}(\tau(a)) \geq 0$.

Of course, one needs to use the Newton formulae (**[Weyl**, p. 38]) to translate the $\tau_j(a)$ into functions of the b_j. For example, $\tau_1 = \sigma_1$, $\tau_2 = \sigma_1^2 - 2\sigma_2$, $\tau_3 = \sigma_1^3 - 3\sigma_1\sigma_2 + 3\sigma_3, \ldots$. When $n = 2$, the condition for all the roots of f to be real is that the matrix

$$B = \begin{pmatrix} 2 & b_1 \\ b_1 & b_1^2 - 2b_2 \end{pmatrix}$$

be positive semidefinite. This is equivalent to the determinant of B being nonnegative, where this determinant is, of course, just the discriminant $b_1^2 - 4b_2$ of f.

(6.7) We sketch a proof of (6.2): Let V denote $W \otimes_{\mathbf{R}} \mathbf{C}$. The bilinear form $(\ ,\ )$ extends naturally to (complex) bilinear forms (denoted $(\ ,\ )$, of course) on V and V^*. Then $G = K_{\mathbf{C}}$ acts on V and V^* and preserves $(\ ,\ )$. The invariants $p_i \in \mathbf{R}[W]^K$ extend naturally to elements (also denoted p_i) in $\mathbf{C}[V]^G$, and $p: V \to Y \subseteq \mathbf{C}^d$ has the properties listed in (3.2) (replacing q by p and e by d). Note that $Z = Y \cap \mathbf{R}^d$. Let $\langle v', v\rangle$ denote $(v', \overline{v})$ for $v, v' \in V$. Then $\langle\ ,\ \rangle$ is a K-invariant hermitian form on V.

Let $z \in Z$. We want to show that $\text{Grad}(z) \geq 0$ implies that $z \in X$, i.e. that $p^{-1}(z)$ contains a real point. Choose $v \in M$ (the Kempf-Ness set of V) such that $p(v) = z$. Then $p(\overline{v}) = \overline{p(v)} = \overline{z} = z$, hence v and $\overline{v}$ are in the same fiber of p. Now v is a closest point to the origin in this fiber and $\|\overline{v}\| = \|v\|$. By (4.2.3) we see that $\overline{v} \in M$ and $\overline{v} = kv$ for some $k \in K$. Write $v = w_1 + iw_2$ where $w_1, w_2 \in W$. Then $kw_1 = w_1$ and $kw_2 = -w_2$.

(6.8) We need a consequence of Luna's slice theorem ([**Lu1**]). Let $x \in V$ and define:

$$\Delta(x) := \{\xi \in (V^*)^{G_x} : \xi \text{ annihilates } T_x(Gx)\}.$$

$$D(x) := \text{span}_{\mathbf{C}}\{dp_i(x)\}.$$

By the chain rule,

$$D(x) = \text{span}_{\mathbf{C}}\{df(x) : f \in \mathbf{C}[V]^G\}.$$

Obviously, $D(x) \subseteq \Delta(x)$.

(6.9) PROPOSITION (see [**PS**]). *Assume that Gx is closed. Then $D(x) = \Delta(x)$.*

(6.10) Let v and k be as in (6.7). Set

$$\Delta_{\mathbf{R}}(v) = \{\xi \in \Delta(v) : \xi \circ k = \overline{\xi}\}.$$

An easy computation ([**PS**]) shows that

(6.11) LEMMA. *$\Delta_{\mathbf{R}}(v)$ is the real span of $\{dp_i(v)\}$.*

PROOF OF (6.2): Suppose that $\text{Grad}(z) \geq 0$. Since $\text{Grad}_{ij} = (dp_i, dp_j)$, (6.11) shows that $(\ ,\ ) \geq 0$ on $\Delta_{\mathbf{R}}(v)$. Define $\lambda_1 \in V^*$ by $\lambda_1(v') := (v', \overline{v}) = \langle v', v\rangle$, $v' \in V$. Then $\lambda_1 \in \Delta(v)$ since it is K_v-invariant (hence G_v-invariant) and annihilates $T_v(Gv)$ (since $v \in M$). Define λ_2 by $\lambda_2(v') := (v', v)$. Then λ_2 is the differential at v of the invariant $x \mapsto 1/2(x, x)$, hence $\lambda_2 \in \Delta(v)$. The difference $\lambda = \lambda_2 - \lambda_1$ lies in $\Delta_{\mathbf{R}}(v)$ since $\lambda(v') = (v', 2iw_2)$, $v' \in V$. Thus $0 \leq (\lambda, \lambda) = (2iw_2, 2iw_2)$. This forces $w_2 = 0$, hence $v = w_1 \in W$ and $z = p(w_1) \in X$. ∎

§ 7. The Deformation Retraction

(7.0) We finish the proof of Neeman's theorem (5.1). An essential ingredient is an inequality of Łojasiewicz, which we now recount.

Let $f: U \to \mathbf{R}$ be a real analytic function, where U is a neighborhood of $0 \in \mathbf{R}^n$ and $f(0) = 0$. Let $|\ \ |$ denote the usual Euclidean norm.

(7.1) THEOREM (Łojasiewicz (see [**Łoj**], [**BiMl**, Prop. 6.8])). *Let f be as above. Then there is a neighborhood $U' \subseteq U$ of 0 and a constant θ, $0 < \theta < 1$, such that*

$$|(\operatorname{grad} f)(x)| \geq |f(x)|^\theta, \qquad x \in U'.$$

(7.2) COROLLARY. *Let f be a homogeneous polynomial function of degree $d > 0$. Then there are constants C, θ with $C > 0$ and $0 < \theta < 1$ such that*

$$|(\operatorname{grad} f)(x)||x|^\delta \geq C|f(x)|^\theta, \qquad x \in \mathbf{R}^n,$$

where $\delta = 1 - d(1 - \theta) \geq 0$.

PROOF: The inequality (7.1) forces the degree of homogeneity of $|f(x)|^\theta$ to be at least that of $|(\operatorname{grad} f)(x)|$, so $\delta \geq 0$. Now restrict the inequality to a small disk about 0 in U'. On the boundary of the disk $|x|$ is constant, so an inequality as written holds there. By homogeneity, the inequality extends radially. ∎

(7.3) EXAMPLE (Bierstone-Kucharz): The smallest possible value of θ in (7.2) is $1 - 1/d$. The inequality may not hold with this value of θ, however: Let $f(x, y, z) = x^2 z - 2xy^2$. Along the curve defined by $z = 1$ and $x = y^2$ we have $|\operatorname{grad} f| = \sqrt{16y^6 + y^8}$ and $|f| = y^4$, so θ must be at least $3/4 > 2/3 = 1 - 1/d$.

(7.4) Let $h(v) = \|f_v\|^2$ as in (5.3), let ψ_t be the local 1-parameter group generated by $-\operatorname{grad} h$, and let $g(x) = \|x\|^2$, $x \in V$. Since h is homogeneous of degree 4, $((-\operatorname{grad} h)g)(v) = -8h(v) \leq 0$, so that the flow ψ_t, $t \geq 0$, preserves the ball B_r of radius r in V for all r. Thus ψ_t exists for all $t \geq 0$.

We will establish the following estimate:

(7.5) LEMMA. *Let $r > 0$. Then $\int_t^\infty \|\frac{d}{ds}\psi_s(v)\|\, ds \to 0$ as $t \to \infty$, uniformly for $v \in B_r$.*

PROOF OF (5.1): Choose a homeomorphism τ of $[0, 1]$ with the one-point compactification $[0, \infty]$ of $[0, \infty)$ so that $\tau([0, 1)) = [0, \infty)$. Define

$\varphi_t := \psi_{\tau(t)}$. It follows from (7.5) that $\varphi_t \colon [0,1) \times B_r \to B_r$ is uniformly continuous, hence it has a continuous extension to $[0,1] \times B_r$. Letting $r \to \infty$ we obtain $\varphi_t \colon [0,1] \times V \to V$ with the desired properties.

PROOF OF (7.5): We consider the function $H(t) := h(\psi_t(v))$. Since $-\operatorname{grad} h$ generates ψ_t, we have

$$H'(t) = -\|\operatorname{grad} h(\psi_t(v))\|^2 = -\|\frac{d}{dt}\psi_t(v)\|^2.$$

We need to show that $\int_t^\infty |H'(s)|^{1/2}\, ds \to 0$ uniformly for $v \in B_r$ as $t \to \infty$. We may assume that $v \notin M$, i.e., we may assume that $H(t)$ never vanishes. Applying (7.2) to h we obtain an inequality

$$|H'(t)|^{1/2}\|\psi_t(v)\|^\delta \geq CH(t)^\theta, \qquad 3/4 \leq \theta < 1,$$

where $\delta = 4\theta - 3$. Along the curve $\psi_t(v)$, $t \geq 0$, the norm of $\psi_t(v)$ is bounded by r, so we obtain an estimate

$$\textbf{(7.6)} \qquad -H'(t) \geq C'H(t)^{2-\epsilon}$$

where C' is a positive constant independent of $v \in B_r$, and $2 - \epsilon = 2\theta$ (so $0 < \epsilon \leq 1/2$). We may rewrite (7.6) as

$$\frac{d}{dt}H(t)^{\epsilon-1} \geq C'(1-\epsilon),$$

and integrating we obtain

$$H(t)^{\epsilon-1} \geq C'(1-\epsilon)t, \qquad t \geq 0.$$

Hence there is an estimate

$$H(t) \leq C''t^{-1/(1-\epsilon)}, \qquad t > 0,$$

where $C'' = (C'(1-\epsilon))^{-1/(1-\epsilon)}$. We are almost done! Let $t > 0$ and set $F(s) = s^{1+\epsilon}$. Then $F'(s) = (1+\epsilon)s^\epsilon$, and $H(s)F'(s) \in L^1([t,\infty])$ since $-1/(1-\epsilon) + \epsilon < -1$. Integration by parts shows that $H'(s)F(s) \in L^1([t,\infty])$ (since $\lim_{s\to\infty} H(s)F(s) = 0$). Now $1/\sqrt{F} \in L^2([t,\infty])$ and $\sqrt{|H'|} \cdot \sqrt{F} \in L^2([t,\infty])$, hence the product $\sqrt{|H'|} \in L^1([t,\infty])$. The L^1-norm of $\sqrt{|H'|}$ is clearly bounded by a function of t which tends to 0 as $t \to \infty$, uniformly for $v \in B_r$. ∎

References

[**BiMl**] E. Bierstone and P. Milman, *Semianalytic and subanalytic sets*, Publ. Math. IHES. **67** (1988), 5–42.

[**Br**] G. Bredon, "Introduction to Compact Transformation Groups," Academic Press, New York, 1972.

[**Chev**] C. Chevalley, "Theory of Lie Groups," Princeton University Press, Princeton, 1946.

[**DaK**] J. Dadok and V. Kac, *Polar representations*, J. Alg. **92** (1985), 504–524.

[**GR**] R. C. Gunning and H. Rossi, "Analytic Functions of Several Complex Variables," Prentice Hall, Englewood Cliffs, 1965.

[**Helg**] S. Helgason, "Differential Geometry, Lie Groups, and Symmetric Spaces," Academic Press, New York, 1978.

[**Hoch**] G. Hochschild, "The Structure of Lie Groups," Holden-Day, San Francisco, 1965.

[**Humph**] J. E. Humphreys, "Linear Algebraic Groups," Graduate Texts in Mathematics vol. **21**, Springer-Verlag, Berlin Heidelberg New York, 1975.

[**KN**] G. Kempf and L. Ness, *The length of vectors in representation spaces*, in "Algebraic Geometry," Lect. Notes Math. vol. **732**, Springer-Verlag, Berlin Heidelberg New York, 1979, pp. 233–243.

[**Kr**] H. Kraft, "Geometrische Methoden in der Invariantentheorie," Vieweg-Verlag, Braunschweig, 1984.

[**KPR**] H. Kraft, T. Petrie and J. Randall, *Quotient varieties*, Adv. Math. (to appear).

[**Łoj**] S. Łojasiewicz, "Ensembles semi-analytiques," IHES-notes, Bures-sur-Yvette, 1965.

[**Lu1**] D. Luna, *Slices étales*, Bull. Soc. Math. France, Mémoire **33** (1973), 81–105.

[**Lu2**] ———, *Sur certaines opérations différentiables des groupes de Lie*, Amer. J. Math. **97** (1975), 172–181.

[**Lu3**] ———, *Fonctions différentiables invariantes sous l'opération d'un groupe réductif*, Ann. Inst. Fourier **26** (1976), 33–49.

[**MumF**] D. Mumford and J. Fogarty, "Geometric Invariant Theory," 2nd edition, Springer-Verlag, Berlin Heidelberg New York, 1982.

[**Ne**] A. Neeman, *The topology of quotient varieties*, Ann. of Math. **103** (1985), 419–459.

[**Ol**] R. Oliver, *A proof of the Connor Conjecture*, Ann. of Math. **103** (1976), 637–644.

[**PS**] C. Procesi and G. Schwarz, *Inequalities defining orbit spaces*, Inv. Math. **81** (1985), 539–554.

[**Schw**] G. Schwarz, *Smooth functions invariant under the action of a compact Lie group*, Topology **14** (1975), 63–68.

[**Sl**] P. Slodowy, *Some remarks on actions of real reductive groups*, (preprint).

[**Weyl**] H. Weyl, "The Classical Groups," 2nd ed., Princeton University Press, Princeton, 1946.

[**Wh**] H. Whitney, "Complex Analytic Varieties," Addison-Wesley, Reading, 1972.

Gerald W. Schwarz
Department of Mathematics
Brandeis University
PO Box 9110
Waltham, MA 02254-9110
USA

RATIONALITY OF MODULI SPACES VIA INVARIANT THEORY

Nicholas I. Shepherd-Barron

The aim of this paper is to convey some idea, by means of concrete examples, of what it means for a moduli space in algebraic geometry to be rational, and of what this question has to do with group theory.

For a thorough explanation of the connection with group theory, see [**15**]. Here, however, we just describe a couple of moduli spaces (namely for curves of genus 1 and 2) which are very well known, and then discuss some approaches to the question of whether quotient spaces $\mathbf{P}/G$ of projective spaces by reductive groups are rational. The connection between these two problems is that moduli spaces are sometimes describable as such quotients. For details and further information, see Dolgachev's survey paper [**8**] and the other references below.

We begin with some basic definitions, working throughout over the field $\mathbf{C}$ of complex numbers. Two complex algebraic varieties X and Y are *birational* (denoted $X \simeq Y$) if their function fields $\mathbf{C}(X)$ and $\mathbf{C}(Y)$ are isomorphic (here, "function" means "meromorphic complex-valued function"). That is X and Y have the same function-theory. More geometrically, this means that there are dense open (in the Zariski topology) subvarieties X_0 of X and Y_0 of Y with X_0 isomorphic to Y_0. An n-dimensional variety V is *rational* if it is birational to projective space $\mathbf{P}^n$.

If V parametrizes a family $\{X_t\}_{t\in V}$ of geometrical objects and $\dim V = n$, then to say that V is rational means that we can write down equations defining almost all the members X_t of the family, where the coefficients of the equations depend on exactly n free parameters, and such that the generic X_t appears exactly once as the parameters vary.

Recall that the moduli space M_g for smooth projective curves (= compact Riemann surfaces) of genus g is an algebraic variety whose points classify the isomorphism classes of all these curves. (There are some other requirements that we shall ignore.) It is a theorem that M_g is a quasi-projective variety [**15**].

For example, consider the moduli space M_1 for elliptic curves (i.e., compact Riemann surfaces of genus 1). Any elliptic curve C is a complex Lie group, and so the automorphism group $\operatorname{Aut} C$ acts transitively. Let $x \in C$, and consider the vector space $H^0(\mathcal{O}_C(2x))$ of meromorphic functions on C with at most a double pole at x and holomorphic away from x. This is two-dimensional, as shown, for example by the Riemann-Roch theorem. If we write $C = \mathbf{C}/\Lambda$, with $\Lambda \cong \mathbf{Z} \oplus \mathbf{Z}$ a lattice, and if $x = 0$, then $\{1, \wp\}$ is a basis of $H^0(\mathcal{O}_C(2x))$, where $\wp$ is the Weierstrass $\wp$-function. We get a morphism $\pi: C \to \mathbf{P}^1$ given by $\pi(z) = (1, \wp(z))$ which is of degree 2, and since $\wp'$ is meromorphic but not in $H^0(\mathcal{O}_C(2P))$, the differential equation

$$\wp'^2 = 4\wp^2 - g_2\wp - g_3$$

is an equation describing C (i.e., the function field $\mathbf{C}(C)$ is generated by $\wp$ and $\wp'$ subject to this single relation). Write $F(\wp) = 4\wp^3 - g_2 \cdot \wp - g_3$; then the branch locus of π is {zero-locus of F} $\cup$ $\{\infty\}$. Since $\operatorname{Aut} C$ acts transitively all choices of x are equivalent. So if C' is another elliptic curve and $\pi': C' \to \mathbf{P}^1$ is constructed similarly to π, then $C \cong C'$ if and only if the branch loci of π and π' are equivalent under $\operatorname{Aut} \mathbf{P}^1 = \mathrm{PGL}_2$. Since in addition, given four distinct points $p_1, \ldots, p_4$ in $\mathbf{P}^1$ we can construct a map $\pi: C \to \mathbf{P}^1$ that is of degree 2, with branch locus $\{p_1, \ldots, p_4\}$ and C an elliptic curve, we see that, as a set,

$$\begin{aligned} M_1 &\cong \{\text{unordered 4-tuples of distinct points in } \mathbf{P}^1\}/\,\mathrm{PGL}_2 \\ &\cong \{\text{homogeneous quartic polynomials in 2 variables} \\ &\qquad \text{with no repeated root, modulo scalar multiplication}\}/\,\mathrm{PGL}_2. \\ &\cong \{\text{binary quartics with no repeated root, modulo scalars}\}/\,\mathrm{PGL}_2\,. \end{aligned}$$

Suppose that $F = \Sigma\binom{4}{j} a_j X_1^{4-j} X_2^j$ is a binary quartic. It has long been known [**17**] that the ring of SL_2-invariants for binary quartics is a polynomial ring $\mathbf{C}[S,T]$, where $S = a_0a_4 - 4a_1a_3 + 3a_2^2$ and $T = a_0a_2a_4 + 2a_1a_2a_3 - a_0a_3^2 - a_1^2a_4 - a_2^3$. If now F_1, F_2 are binary quartics, at least one of which has no multiple root, then they are PGL_2-equivalent if and only if there exists $\lambda \in \mathbf{C}^*$ such that $(S(F_2), T(F_2)) = (\lambda^2 S(F_1), \lambda^3 T(F_1))$ (recall that a binary quartic is stable, in Mumford's sense, if and only if it has no multiple roots).

Now compute S, T for $F = 4X_1^3X_2 - g_2X_1X_2^3 - g_3X_2^4$ (a homogenized form of $4\wp^3 - g_2\wp - g_3$); we get $S = g_2$, $T = g_3$. Hence if $\mathbf{A}^2 = \operatorname{Spec} \mathbf{C}[g_2, g_3]$ and $\Delta = g_2^3 - 27g_3^2$ is the discriminant, and if $\mathbf{C}^*$ acts on $\mathbf{A}^2$ by $\lambda(g_2, g_3) = (\lambda^2 g_2, \lambda^3 g_3)$, then the moduli space $M_1 = (\mathbf{A}^2 \setminus \Delta)/\mathbf{C}^*$.

There is a birational slice to this $\mathbf{C}^*$-action given by the equation $g_2 = g_3 = g$, say; then the family $y^2 = 4x^2 - gx - g$ is a family of algebraic varieties parametrized by the affine line $\mathbf{A}^1 = \operatorname{Spec} \mathbf{C}[g]$, and almost every elliptic curve appears exactly once. In fact, of course, $M_1 \cong \operatorname{Spec} \mathbf{C}[j]$, where $j = 1728g_2^3/\Delta = 1728g/(g-27)$. However, there is no universal family over M_1.

The case $g = 2$.

Let C be a curve of genus 2, and $\{\omega_1, \omega_2\}$ a basis for its space $H^0(\omega_C)$ of holomorphic 1-forms. The morphism $\varphi: C \to \mathbf{P}^1$ given by $\varphi(z) = (\omega_1(z), \omega_2(z))$ is of degree 2 and branched at 6 points. Conversely, given $p_1, \ldots, p_6 \in \mathbf{P}^1$, we can construct a curve C of genus 2 and a double cover $\varphi: C \to \mathbf{P}^1$ branched at $p_1, \ldots, p_6$ such that φ is the map associated to a choice of basis of $H^0(\omega_C)$.

Let $\mathcal{S}_6$ denote the symmetric group of 6 letters. As for the case of elliptic curves, we see that

$$(\mathbf{P}^1)^6/\mathcal{S}_6 = \mathbf{P}^6 = \mathbf{P}(\operatorname{Sym}^6 \mathbf{C}^{2*}), \text{ and that } M_2 = \mathbf{P}^6/\operatorname{SL}_2.$$

In fact, if $\mathbf{P}_0^6$ is the subspace of $\mathbf{P}^6$ consisting of sextic polynomials with no multiple root, then $M_2 = \mathbf{P}_0^6/\operatorname{SL}_2$ (which does exist as a geometric quotient). To describe this, we examine the ring R of SL_2-invariants for binary sextics. Clebsch proved that $R = \mathbf{C}[A_2, B_4, C_6, D_{10}, E_{15}]$, where subscripts denote degrees in the coefficients $a_0, \ldots, a_6$ of our binary sextic, and the only relation is $E^2 = P_{30}(A, B, C, D)$ for an appropriate P [**17**]. We can take D to be the discriminant; then

$$\begin{aligned} \mathbf{P}_0^6/\operatorname{SL}_2 &\cong \operatorname{Proj} R \setminus \{D = 0\} \cong \operatorname{Proj} R^{(2)} \setminus \{D = 0\} \\ &\cong \operatorname{Proj} \mathbf{C}[A, B, C, D] \setminus \{D = 0\}. \end{aligned}$$

(Recall that if $R = \oplus R_n$ is a graded ring, then $R^{(2)} = \oplus R_{2n}$, and $\operatorname{Proj} R = \operatorname{Proj} R^{(2)}$). This last space is trivially rational, and is isomorphic to $\mathbf{A}^3/H$, where $H = \langle \sigma \rangle$ is a group of order 5 acting via $\sigma(x, y, z) = (\varepsilon x, \varepsilon^2 y, \varepsilon^3 z)$, $\varepsilon = \exp(2\pi i/5)$. From this, we see also that M_2 is contractible.

Clebsch's proof of the structure of R is hard, and so we sketch here another proof that M_2 is rational, due to Coble [**7**].

Put $(\mathbf{P}^1)_0^6 = \{(x_1, \ldots, x_6) \in (\mathbf{P}^1)^6 |$ all x_i are distinct$\}$. Then SL_2 acts diagonally on $(\mathbf{P}^1)_0^6$, and a geometric quotient $(\mathbf{P}^1)_0^6/\operatorname{SL}_2 = X$, say, exists. Take $p_1, \ldots, p_5 \in \mathbf{P}^3$ in general position; we may assume that $p_1 = (1,0,0,0), \ldots, p_4 = (0,0,0,1)$ and $p_5 = (1,1,1,1)$. Put $L_{ijk} = \langle p_i, p_j, p_k \rangle$, the 2-plane spanned by p_i, p_j, p_k, and put $Y = \mathbf{P}^3 \setminus \bigcup_{i,j,k} L_{ijk}$. Then there is an isomorphism $\varphi: Y \xrightarrow{\sim} X$ obtained as follows: Let $y \in Y$. Then there is a unique twisted cubic Γ through $p_1, \dots, p_5, y$; via the choice of an isomorphism $\Gamma \xrightarrow{\sim} \mathbf{P}^1$ we define $\varphi(y)$ as the SL_2-equivalence class of $(p_1, \dots, p_5, y)$.

Now the symmetric group $\mathcal{S}_6$ acts on X, via the permutation action on $(\mathbf{P}^1)^6$, and clearly $M_2 = X/\mathcal{S}_6$. Hence $M_2 = /\mathcal{S}_6$; we can describe the action of $\mathcal{S}_6$ on Y as follows. The group $\mathcal{S}_5$ acts linearly on $\mathbf{P}^3$ so as to permute $\{p_1, \dots, p_5\}$. Let τ be the Cremona transformation $(x_0, \dots, x_3) \mapsto (x_0{}^{-1}, \dots, x_3{}^{-1})$; then $\langle \tau, \mathcal{S}_5 \rangle = G$ is a group of Cremona transformations isomorphic to $\mathcal{S}_6$. Let $Y_1 \to \mathbf{P}^3$ be the blow-up along $p_1, \dots, p_5$. The linear system $\left|2H - \sum_{i=1}^5 p_i\right|$ of quadrics through $p_1, \dots, p_5$ defines a morphism $\varphi: Y_1 \to \mathbf{P}^4$ that is birational onto its image Z, the *Segre cubic*. Z has 10 nodes, the images of (the strict transforms of) the lines $\langle p_i, p_j \rangle$, and 15 planes, the images of the planes L_{ijk} and the points p_i. For us, the point is that now $\mathcal{S}_6$ acts biregularly on Z, and this action is induced from a linear action of $\mathcal{S}_6$ on $\mathbf{P}^4$. That is, $\mathbf{P}^4 = \mathbf{P}(V)$ as $\mathcal{S}_6$-spaces, where V is a 5-dimensional representation of $\mathcal{S}_6$ which is equivalent, via an outer automorphism of $\mathcal{S}_6$, to $V_1 \otimes \sigma$, where V_1 is the representation of S_6 as the Weyl group $W(\mathcal{A}_5)$ and σ is the signature; i.e., $1 \oplus V_1$ is the standard permutation representation. If σ_α is the α'th elementary symmetric polynomial in six variables and δ_{15} is the square root of the discriminant, then Z, as a subvariety of $\mathbf{P}^5 = \mathbf{P}(\sigma \otimes (V \oplus 1))$, is given by the equation $\sigma_1 = \sigma_3 = 0$, and so

$$\Sigma/S_6 \cong \operatorname{Proj} \mathbf{C}[\sigma_2, \sigma_4, \sigma_5^2, \sigma_6, \delta_{15}] = \operatorname{Proj} \mathbf{C}[\sigma_2, \sigma_4, \sigma_5^2, \sigma_6],$$

which is certainly rational. Hence again M_2 is rational.

Remarks.

(1) The identification $X = Y$ above of course goes through for ordered n-tuples of points on $\mathbf{P}^1$, for any n. In fact, it seems very likely that if $p_1, \dots, p_{n+2} \in \mathbf{P}^n$ are points in general position and if $\widetilde{\mathbf{P}} \to \mathbf{P}^n$ is obtained by successively blowing up points, lines, planes, ..., spanned by the various subsets of $\{p_1, \dots, p_{n+2}\}$, then $\widetilde{\mathbf{P}}$ is isomorphic to $M_{0,n+3}$, the moduli space for "stable n-pointed curves of genus zero", and the smooth fibres of the natural projection $M_{0,n+3} \to M_{0,n+2}$ (identifying $M_{0,n+3}$ with the universal curve over $M_{0,n+2}$) are the strict transforms of the rational normal n-ics through $p_1, \dots, p_{n+2}$. However, I do not know a proof of this. Unfortunately, where for $n = 3$ we were led to the Segre cubic, we find no such simple variety for $n \geq 4$.

(2) The family of twisted cubics through $p_1, \dots, p_5$ in $\mathbf{P}^3$ can be used to give another proof of a theorem of Maeda (as yet unpublished) that $\mathbf{P}^3/\mathcal{A}_5$

is rational, where $\mathbf{P}^3 = \mathbf{P}(V)$ and $1 \oplus V$ is the permutation representation of the alternating group $\mathcal{A}_5$, as follows.

Let $f: \widetilde{\mathbf{P}} \to \mathbf{P}^3$ be as in (1). The system $\left|5H - 3 \cdot \sum_{i=1}^{5} p_i\right|$ defines a morphism $g: \widetilde{\mathbf{P}} \to \mathbf{P}^5$ whose image is a quintic Del Pezzo surface $F (\cong M_{0,5})$. The smooth fibres of g are the strict transforms of the twisted cubics through $p_1, \ldots, p_5$. The sheaf $\mathcal{L} = \pi^*\mathcal{O}(1)$ is $\mathcal{A}_5$-linearized and cuts out $\mathcal{O}(3)$ on $g^{-1}(x) \cong \mathbf{P}^1$. Let F_0 be the subvariety of F where A_5 acts freely, and put $\widetilde{P}_0 = g^{-1}(F_0)$. Then $\mathcal{L}$ descends to an invertible sheaf $\mathcal{M}$ on $\widetilde{P}_0/\mathcal{A}_5$ that cuts out $\mathcal{O}(3)$ on the generic fibre of $h: \widetilde{P}_0/\mathcal{A}_5 \to F_0/\mathcal{A}_5$. Let ω be the relative dualizing sheaf of h; then $\mathcal{M} \otimes \omega$ cuts out $\mathcal{O}(1)$ on the generic fibre of h, and so the Severi-Brauer scheme h is trivial. Hence $\widetilde{\mathbf{P}}_0/\mathcal{A}_5 \simeq (F_0/\mathcal{A}_5) \times \mathbf{P}^1$, and $F_0/\mathcal{A}_5$ is rational by Castelnuovo's criterion.

In fact, $\widetilde{\mathbf{P}}_0/\mathcal{A}_5$ is rational over $\mathbf{Q}$. It is enough to show that $F_0/\mathcal{A}_5$ is rational over $\mathbf{Q}$, where F is obtained by blowing up four $\mathbf{Q}$-rational points in general position in $\mathbf{P}^2$. Then $F \cong M_{0,5}$ and in turn $M_{0,5}/\mathcal{S}_5 \cong (\mathbf{P}^5)^s/\operatorname{SL}_2$, $\mathbf{P}^5$ being the space of binary quintics, and $M_{0,5}/\mathcal{A}_5$ is the double cover of $M_{0,5}/\mathcal{S}_5$ branched along the discriminant locus. By classical invariant theory, $(\mathbf{P}^5)^s/\operatorname{SL}_2 = \operatorname{Proj} \mathbf{C}[A_4, B_8, C_{12}, D_{18}]$, where B can be taken to be the discriminant, and the only relation is $D^2 = P_{36}(A, B, C)$. Then

$$
\begin{aligned}
F/\mathcal{A}_5 &\cong \operatorname{Proj}(\mathbf{C}[A, B, C, D, \delta_4]/(D^2 - P(A, B, C), \delta^2 - B)) \\
&\cong \operatorname{Proj} \mathbf{C}[A, C, \delta],
\end{aligned}
$$

which is rational.

(3) In contrast to the case $g = 1$, there is no family parametrized by any part of M_2 that contains almost all curves of genus two exactly once. A similar thing happens for hyperelliptic curves in any even genus.

The case $g = 3$.

Let C be a curve of genus 3, and $\{\omega_1, \omega_2, \omega_3\}$ a basis of $H^0(\omega_C)$. Then the map $f: C \to \mathbf{P}^2$ given by $f(z) = (\omega_1(z), \omega_2(z), \omega_3(z))$ is either 2-to-1 onto a conic (when C is hyperelliptic) or an isomorphism onto a quartic (this is the generic case). Hence M_3 is birational to $\mathbf{P}^{14}/\operatorname{SL}_3$, where $\mathbf{P}^{14}$ is the projectivized space of homogeneous quartic polynomials in three variables ("ternary quartics"). It is still not known whether M_3 is rational. Although one can show that $M_3 \times \mathbf{P}^2$ is rational, this is not enough; there exist irrational 3-folds X such that $X \times \mathbf{P}^3$ is rational (so that X is *stably rational*) [1].

The cases $g = 4, 5, 6$.

For $4 \leq g \leq 6$, it turns out that as for $g \leq 3$, M_g is birational to some quotient $\mathbf{P}/G$, where the reductive group G acts linearly on the projective space $\mathbf{P}$. For $g = 4$ and 6, M_g is rational, while the rationality of M_5 is unknown.

The cases $g \geq 7$.

For $g \geq 23$, M_g is not even unirational, i.e., there is no generically surjective rational map $\mathbf{P}^N \to M_g$ ([**9**],[**11**], [**12**]), and so in particular cannot be birationally equivalent to a quotient $\mathbf{P}/G$. For $7 \leq g \leq 13$, M_g is unirational [**6**], but no description as a quotient $\mathbf{P}/G$ is known.

Description of $\mathbf{P}(V)/G$.

Suppose that V is a representation of the reductive group G. By Mumford's results [**15**], there is an open subvariety $\mathbf{P}^s$ of $\mathbf{P} = \mathbf{P}(V)$, the set of *stable points*, such that $\mathbf{P}^s/G$ exists as a quasi-projective geometric quotient (in particular, the points of $\mathbf{P}^s/G$ correspond to G-orbits in $\mathbf{P}$). There is also an open subvariety $\mathbf{P}^{ss}$, with $\mathbf{P}^s \subseteq \mathbf{P}^{ss} \subseteq \mathbf{P}$, the set of *semi-stable points*, such that the set of closed G-orbits in $\mathbf{P}^{ss}$ forms a projective variety $\mathbf{P}^{ss}/G$. If $R = \mathbf{C}[V^*]$, the homogeneous co-ordinate ring of $\mathbf{P}$, so that $\mathbf{P} = \operatorname{Proj} R$, then $\mathbf{P}^{ss}/\!\!/G = \operatorname{Proj} R^G$, where R^G is the subring of R consisting of the G-invariants.

Classically, invariant theory aimed to describe R^G in terms of generators and relations, or equivalently, to describe $\mathbf{P}^{ss}/\!\!/G$ via co-ordinates and equations. However, this description can easily be rather complex even when $\mathbf{P}^{ss}/\!\!/G$ has a straightforward description in other terms. For example, take G to be SL_2 and V the space of $2 \times n$ matrices on which G acts by left multiplication. Then $\mathbf{P}^s = \mathbf{P}^{ss}$ consists of the matrices of rank 2 and $\mathbf{P}^s/G = \mathrm{Gr}(2,n)$, a Grassmannian. The ring of invariants, however, has $\binom{n}{2}$ generators (the Plücker co-ordinates) and $\binom{n}{4}$ relations (the Plücker equations); this complexity means that the Grassmannian is not necessarily best studied from this point of view.

This complexity is really what motivates the search for a simple birational description of $\mathbf{P}^s/G$. Concretely, we ask whether $\mathbf{P}^s/G$ is rational. Rather more vaguely, given that classifying all the stable G-orbits in $\mathbf{P}$ may be very involved, we ask whether it is possible to classify "most" orbits simply.

In general, $\mathbf{P}^s/G$ need not be rational; Saltman has shown [**18,19**] that for various finite groups G, $\mathbf{P}(V)/G$ is never stably rational if V is a faithful representation. On the other hand, the quotients most often arising in algebraic geometry, say as moduli spaces, are often easily seen to be stably rational (for example, by the "no-name" method: see [**8**,§4]). Our aim here, then, is to discuss some techniques for establishing the rationality of quotient varieties, concentrating on those that are not covered in Dolgachev's paper [**8**]; our knowledge of higher-dimensional algebraic geometry is not enough for anything deeper.

Techniques.

These are the variations on the following theme. Suppose that $\varphi: X \to Y$ is a dominant projective G-equivariant morphism of G-varieties. Assume that there is a G-equivariant embedding $Y \hookrightarrow \mathbf{P}(V)$ such that V is a representation of G and $Y_0 = Y \cap \mathbf{P}(V)^{vs}$ is non-empty $(\mathbf{P}(v))^{vs}$ denotes the set of stable points whose stabilizer is the kernel Z of the G-action on $\mathbf{P}(V)$), so that by Luna's étale slice theorem, the map $Y_0 \to Y_0/G$ is a principal (G/Z)-bundle in the étale topology. (The étale slice theorem [**15**, pp. 152–153] is usually stated for affine varieties. However, Y_0 is covered by G-stable affine charts.) Moreover, putting $X_0 = \varphi^{-1}(Y_0)$, there is a Cartesian diagram,

$$\begin{array}{ccc} X_0 & \longrightarrow & X_0/G \\ {\scriptstyle\varphi}\downarrow & & \downarrow{\scriptstyle\psi} \\ Y_0 & \longrightarrow & Y_0/G \end{array}$$

and the geometric fibres of ψ are isomorphic to those of φ. If φ is G-equivariantly birational to a bundle with rational fibre, then one can hope that ψ is also birational to a bundle, so that $X_0/G \simeq (Y_0/G) \times \mathbf{P}^n$. Specifically, if ψ is a projective space bundle and if there is a (G/Z)-linearized line bundle $\mathcal{L}$ on X or X_0 [**15**, § 1.3] whose restriction to a fibre of φ is $\mathcal{O}(1)$, then $\mathcal{L}$ descends to a line bundle $\mathcal{M}$ on X_0/G that cuts out $\mathcal{O}(1)$ on a fibre of ψ. Then ψ (which is just a Severi-Brauer scheme, a priori) is birationally trivial. (Recall that if $\mathcal{L}$ is very ample, then $\mathcal{L}$ is (G/Z)-linearized if and only if it is G-linearized and $H^0(\mathcal{L})$ is a representation of G/Z and not only of G.)

Example (the no-name method). *Suppose that U, V, W are representations of G, of dimensions m, n, p respectively, that U^{vs}, and V^{vs} are non-empty and that $p \geq m$. Assume that U/G is rational. Then so are $(V/G) \times \mathbf{C}^m$ and $(V \oplus W)/G$.*

PROOF: Regard $U \oplus W$ as a G-linearized vector bundle over U and over V, via the two projections. Then by the results from descent theory mentioned above, $(U \oplus V)/G$ is generically a vector bundle over U/G and over V/G. Hence

$$(V/G) \times \mathbf{C}^m \simeq (U \oplus V)/G \simeq (U/G) \times \mathbf{C}^n,$$

which is rational. Similarly,

$$(V \oplus W)/G \simeq (V/G) \times \mathbf{C}^p \simeq ((V/G) \times \mathbf{C}^m) \times \mathbf{C}^{p-m},$$

and so is rational.

So, in particular, the representation $V \oplus W$ is easier to deal with than either of its factors. The no-name method can be extended, e.g., to products of projective spaces.

Example (a special case of the no-name method).

Let $\mathbf{P}^n$ denote the projectivized space of binary n-ics, $X = \mathbf{P}^8 \times \mathbf{P}^6$, $Y = \mathbf{P}^8$, $\varphi: X \to Y$ the first projection and $G = \mathrm{PGL}_2$. Then X/G is birationally a $\mathbf{P}^8$-bundle over Y/G.

To see concretely that $\psi: X/G \to Y/G$ is a trivial Severi-Brauer scheme, one needs to know that Y^{vs} is non-empty and to find an invariant F of the G-action on X that is linear in the $\mathbf{P}^8$-factor; then the zero-locus of F cuts out a hyper-plane on the generic fibre of ψ, thus trivializing the Severi-Brauer scheme. To find such an F, take $f = \sum \binom{8}{i} \alpha_i x_1^{8-i} x_2^i \in \mathbf{P}^8$ and $g = \sum \binom{6}{j} \beta_j x_1^{6-j} x_2^j \in \mathbf{P}^6$, so that $(\alpha_0, \dots, \alpha_8; \beta_0, \dots, \beta_6)$ are bihomogeneous co-ordinates on X. We can write down F using the "symbolical method" (for a comprehensible account of this, see [**10**]). Symbolically, write $f = A_x^8$, $g = a_x^6 = b_x^6$, where $A_x = A_1 x_1 + A_2 x_2$, etc. $(Aa) = A_1 a_2 - A_2 a_1$, etc. and $A_1^{8-i} A_2^i = \alpha_i$, $a_1^{6-j} a_2^j = b_1^{6-j} b_2^j = \beta_j$. Take $F = (Aa)^4 (Ab)^4 (ab)^2$; this will suffice provided that it is not identically zero. To check this, note that the coefficient of $A_2^8 = \alpha_8$ is

$$\begin{aligned} a_1^4 b_1^4 (ab)^2 &= a_1^4 b_1^4 (a_1^2 b_2^2 - 2 a_1 a_2 b_1 b_2 + a_2^2 b_1^2) \\ &= \beta_0 \beta_2 - 2\beta_1^2 + \beta_2 \beta_0 = 2(\beta_0 \beta_2 - \beta_1^2). \end{aligned}$$

Another way of proving the existence of F is to note that if $V_d = \mathrm{Sym}^d(\mathbf{C}^2)$ is the space of binary d-ics, then $V_8 \otimes \mathrm{Sym}^2(V_6)$ contains a copy of the trivial representation V_0. This in turn follows from the Clebsch-Gordan decomposition

$$\mathrm{Sym}^2(V_6) = V_{12} \oplus V_8 \oplus V_4 \oplus V_0.$$

This argument of course ignores history, since such decompositions of symmetric and tensor products were first found using symbolical method or a method closely allied to it. More importantly, whereas the Killing-Cartan-Weyl classification of representations by highest weights gives an abstract decomposition of tensor products, the symbolical method, when it applies, gives a concrete description of the decomposition in terms of explicit formulae involving coefficients. In applications to invariant theory, such as those we are dicussing or to which we refer, having such formulae available is frequently crucial in order to prove that some construction or morphism is non-degenerate in an appropriate sense.

Double fibration method [4].

If U, V, W are representations of G such that $\dim U = \dim V + 1$, $\dim W \geq \dim U$ and $W \subset \operatorname{Hom}(U, V)$, then the rational map

$$\pi\colon \mathbf{P}(\operatorname{Hom}(U,V)) \to \mathbf{P}(U) \text{ given by } \pi(\varphi) = \ker\varphi$$

induces a dominant G-equivariant rational map $\sigma\colon \mathbf{P}(W) \to \mathbf{P}(U)$ with linear fibres, provided that W is non-degenerate, i.e., in sufficiently general position in $\operatorname{Hom}(U, V)$. If W is non-degenerate, then σ is generically a projective space bundle, and we can apply the results above. Of course, checking that W is non-degenerate is not necessarily trivial.

Bogomolov and Katsylo use this technique to show that $\mathbf{P}^{2n}/\operatorname{SL}_2$ (and so the moduli space H_{n-1} for hyperelliptic curves of genus $n-1$) is rational, for any n, as follows. (Here, $\mathbf{P}^d = \mathbf{P}(V(d))$, where $V(d) = \operatorname{Sym}^d(\mathbf{C}^2)^*$ is the space of binary d-ics.) For example, take $W = V(6p+10)$, $U = V(2p+6) \oplus V(2p+2)$, $V = V(4p+8)$. Then there is an embedding $W \hookrightarrow \operatorname{Hom}(U, V)$ given symbolically by

$$a_x^{6p+10}(b_x^{2p+6}, c_x^{2p+2}) = (ab)^{2p+4} a_x^{4p+6} b_x^2 + (ac)^{2p+2} a_x^{4p+8},$$

and a small amount of calculation shows that W is non-degenerate. The cases $W = V(6p)$, $W = V(6p+2)$ are treated similarly.

The 2-form trick [21].

If $\dim V$ is odd and $W \to \Lambda^2 V^*$, then the rational map $\pi\colon \mathbf{P}(\Lambda^2 V^*) \to \mathbf{P}(V)\colon \pi(\varphi) = \ker\varphi$ restricts to give a dominant map $\sigma\colon \mathbf{P}(W) \to \mathbf{P}(V)$ that is generically a projective space bundle, provided that W is non-degenerate in a suitable sense. For example, taking $G = \operatorname{SL}_3$, V the representation corresponding to the diagram $\overset{1}{\bullet}\!\!-\!\!\!-\!\!\overset{2n}{\bullet}$ and $W = \operatorname{Sym}^{4n+1}(\mathbf{C}^3)^*$, one can prove that $\mathbf{P}(W)/G$, the moduli space for plane curves of degree $4n+1$, is rational.

When the fibres of φ are non-linear.

One can prove rationality results using a morphism $\varphi\colon X \to Y$ even if the fibres are non-linear. For example, the generic fibre of φ might be an N-dimensional intersection of two quadrics, containing a line L; if some such L can be defined over the function field $\mathbf{C}(Y/G)$ of Y/G, then projection from L shows that $X/G \simeq Y/G \times \mathbf{P}^N$ (again subject to some non-degeneracy requirements). A variant of this applies when $G = \operatorname{SL}_3$, with $X = \mathbf{P}(\operatorname{Sym}^{9d+1}(\mathbf{C}^3)^*)$, $Y = \mathbf{P}(\operatorname{Sym}^4(\mathbf{C}^3)^*)$ and φ is given symbolically by

$$\varphi(a_x^4) = (abc)^{3d}(abd)^{3d}(acd)^{3d}(bcd)^{3d} a_x b_x c_x d_x;$$

it turns out that if $d \geq 2$, then φ is dominant and its generic fibre is a component of a complete intersection of 14 quartics all of which are triple along a linear space L of dimension ≥ 13; L can be defined over $\mathbf{C}(Y/G)$ and is suitably non-degenerate [**21**].

Let us return to the dual fibration and 2-form tricks. In each case, there is

(i) a reductive group H containing G;
(ii) a projective space bundle $\beta: E \to Y$ that is homogeneous with respect to G;
(iii) a birational H-equivariant collapsing $\alpha: E \to \mathbf{P}^N$ such that the fibres of β map to linear subspaces of $\mathbf{P}^N$;
(iv) a linear sub-G-space $\mathbf{P}(W) \hookrightarrow \mathbf{P}^N$ such that the induced rational map $\gamma: \mathbf{P}(W) \to Y$ is dominant, so that γ is G-equivariant and is generically a projective space bundle.

From these we deduce, with suitable hypotheses on the stabilizers in G of a generic point of Y and the existence of a linearized line bundle, that $\mathbf{P}(W)/G$ is birationally equivalent to $Y/G \times \mathbf{P}^m$, for some m.

There are two other kind of examples of this.

(a) Take U, V to be representations of G with $\dim U = m+n$, $\dim V = n$. Take $H = \mathrm{SL}_{m+n} \times \mathrm{SL}_n$. Then there is an H-homogeneous $\mathbf{P}^{n^2-1}$-bundle $E \to \mathrm{Gr}(m, U)$ that collapses birationally to $\mathbf{P}^N = \mathbf{P}(\mathrm{Hom}(U, V))$; the corresponding rational map $\pi: \mathbf{P}^N \to \mathrm{Gr}(m, U)$ is given by $\pi(\varphi) = \ker \varphi$ for $\varphi \in \mathrm{Hom}(U, V)$. So if $W \subset \mathrm{Hom}(U, V)$ is a suitable non-degenerate linear sub-G-space and $\dim W \geq mn + 1$, we get a G-equivariant dominant rational map that is generically a projective space bundle. From this one might hope to deduce the rationality of $\mathbf{P}(W)/G$ by analogue of the no-name method, proving that $\mathrm{Gr}(m, U)/G$ is stably rational.

(b) Take $H = \mathrm{Spin}(10)$ and $W = \mathbf{C}^{16}$ a half-spin representation of H. Let P be a parabolic subgroup of H such that $H/P = Y$ is an 8-fold quadric. Let $U = \mathbf{C}^8$ be a half-spin representation of the Levi component P_0 of P ($P_0 = \mathrm{Spin}(8)$), and put $E = H \times^P \mathbf{P}(U)$. Then E is a $\mathbf{P}^7$-bundle over Y and there is a birational collapsing $E \to \mathbf{P}(W)$. The resulting rational map $\mathbf{P}(W) \to Y$ is given by the system of quadrics through the 10-fold spinor variety in $\mathbf{P}(W)$.

All of these examples and techniques depend upon the fact that in the representations of $\mathrm{SL}(m+n) \times \mathrm{SL}(n)$ on $\mathrm{Hom}(\mathbf{C}^{m+n}, \mathbf{C}^n)$, $\mathrm{SL}(2n+1)$ on $\Lambda^2 \mathbf{C}^{2n+1}$ and $\mathrm{Spin}(10)$ on $\mathbf{C}^{16}$, every vector is unstable (i.e., the only invariant polynomials are the constants). Now quite generally, if V is a representation of the semi-simple group H, then the null-cone N (i.e. the set of unstable vectors) has a decomposition $N = \coprod S_\beta$, where each S_β

is an open subvariety of a homogeneous vector bundle over some H/P_β (where P_β a parabolic subgroup of H) ([**2**], [**13**], [**14**], [**16**]) whose fibres are linear subspaces of V. So if G is a reductive subgroup of H and W a linear sub-G-space of V contained in N, then one might expect to be able to prove, by projecting $\mathbf{P}(W)$ onto an appropriate H/P_β, that $\mathbf{P}(W)/G$ is birationally equivalent to $(H/P_\beta)/G) \times \mathbf{P}^m$, for some m.

Acknowledgement. This work was partially supported by the NSF.

References

[**1**] A. Beauville, J.-L. Colliot-Thélène, J.-J. Sansuc and Sir Peter Swinnerton-Dyer, *Variétés stablement rationelles non-rationelles*, Ann Math. **121** (1985), 283–318.
[**2**] F. A. Bogomolov, *Holomorphic tensors and vector bundles on projective varieties*, Math. USSR Izvestiya **13** (1979), 499–556.
[**3**] F. A. Bogomolov, *Stable rationality of quotient varieties by simply connected groups*, Mat. Sbornik **130** (1986), 3–17.
[**4**] F. A. Bogomolov and P. I. Katsylo, *Rationality of some quotient varieties*, Mat. Sbornik **126** (1985), 584–589.
[**5**] M. Brion, *Sur l'image de l'application moment*, Springer LNM **1296** (1986), 177–192.
[**6**] M. Chang and Z. Ran, *Unirationality of the moduli spaces of curves of genus 11, 13 (and 12)*, Invent. Math. **76** (1984), 41–54.
[**7**] A. B. Coble, *An application of Moore's cross-ratio group to the solution of the sextic equation*, Trans. A.M.S. **12** (1911), 311–325.
[**8**] I. V. Dolgachev, *Rationality of fields of invariants*, in "Algebraic Geometry," Bowdoin, 1985. A.M.S., 1987
[**9**] D. Eisenbud and J. Harris, *The Kodaira dimension of the moduli space of curves of genus* ≥ 23, Invent. Math. **90** (1987), 359–387.
[**10**] J. H. Grace and W. H. Young, "The algebra of invariants," Chelsea Publ. Company, New York, 1903.
[**11**] J. Harris, *On the Kodaira dimension of the moduli space of curves, II. The even-genus case*, Invent. Math. **75** (1984).
[**12**] J. Harris and D. Mumford, *On the Kodaira dimension of the moduli space of curves*, Invent. Math. **67** (1982), 23–97.
[**13**] G. Kempf, *Instability in invariant theory*, Ann. of Math. **108** (1978), 299–316.
[**14**] F. Kirwan, "Cohomology of quotients in algebraic and symplectic geometry," Mathematical Notes **31**, Princeton Univ. Press, 1984.
[**15**] D. Mumford and J. Fogarty, "Geometric Invariant Theory," 2nd edition, Ergebnisse der Mathematik, Springer, 1985.
[**16**] L. Ness, *A stratification of the null-cone via the moment map*, Am. J. Math. **106** (1984), 1281–1330.

[**17**] G. Salmon, "Lessons introductory to the modern higher algebra," Chelsea Publ. Company, New York.

[**18**] D. Saltman, *Noether's problem over an algebarically closed field*, Invent. Math. **77** (1984), 71–84.

[**19**] D. Saltman, *Multiplicative field invariants*, J. Alg. **106** (1987), 221–238.

[**20**] N. I. Shepherd-Barron, *The rationality of certain spaces associated to trigonal curves*, in "Algebraic Geometry," Bowdoin, 1985. A.M.S. 1987

[**21**] N. I. Shepherd-Barron, *The rationality of some moduli spaces of plane curves*, Comp. Math. **67** (1988), 51–88.

[**22**] N. I. Shepherd-Barron, *Invariant theory for S_5 and the rationality of M_6*, Comp. Math. (to appear).

[**23**] N. I. Shepherd-Barron, *Apolarity and its applications*, unpublished.

Nicholas I. Shepherd-Barron
Department of Mathematics
University of Illinois at Chicago
Chicago, IL 60680
USA

UNIPOTENT ACTIONS ON AFFINE SPACE

DENNIS M. SNOW

Algebraic group actions on affine space, $\mathbf{C}^n$, are determined by finite dimensional algebraic subgroups of the full algebraic automorphism group, $\operatorname{Aut}\mathbf{C}^n$. This group is anti-isomorphic to the group of algebra automorphisms of $F_n = \mathbf{C}[x_1, \dots, x_n]$ by identifying the indeterminates $x_1, \dots, x_n$ with the standard coordinate functions: $\sigma \in \operatorname{Aut}\mathbf{C}^n$ defines $\sigma^* \in \operatorname{Aut} F_n$ by $\sigma^* f(x) = f(\sigma x)$, where $f \in F_n$ and $x \in \mathbf{C}^n$. In fact, an automorphism σ is often defined in terms of the polynomials $(\sigma^* x_1, \dots, \sigma^* x_n)$. Two special subgroups of $\operatorname{Aut}\mathbf{C}^n$ play an important role in determining the structure of the finite dimensional algebraic subgroups. The first is the affine linear group,

$$A_n = \{\sigma = (f_1, \dots, f_n) \in \operatorname{Aut}\mathbf{C}^n \mid \deg f_i \leq 1\}$$

which is the semi-direct product of the general linear group, $GL_n(\mathbf{C})$, and the abelian group of translations, $T_n \cong \mathbf{C}^n$. The second is the 'Jonquière', or 'triangular' subgroup

$$B_n = \{\sigma = (f_1, \dots, f_n) \in \operatorname{Aut}\mathbf{C}^n \mid f_i = c_i x_i + h_i,\ c_i \in \mathbf{C},\ h_i \in \mathbf{C}[x_{i+1}, \dots, x_n]\}.$$

Since every linear algebraic group is a semi-direct product of a reductive group and a unipotent group, a first step in understanding algebraic subgroups of $\operatorname{Aut}\mathbf{C}^n$ is to deal with these two types of groups separately. This leads to the following problems:

Linearization Problem. *Determine which reductive subgroups of* $\operatorname{Aut}\mathbf{C}^n$ *are conjugate to a subgroup of* A_n.

Unipotent Subgroup Problem. *Determine which unipotent algebraic subgroups of* $\operatorname{Aut}\mathbf{C}^n$ *are conjugate to a subgroup of* B_n.

Due to the fact that $\operatorname{Aut}\mathbf{C}^2$ is an amalgamated product of A_2 and B_2 (over $A_2 \cap B_2$), one can show that any algebraic subgroup of $\operatorname{Aut}\mathbf{C}^2$

is conjugate to a subgroup of A_2 or B_2 with reductive groups going to A_2 and unipotent groups to B_2, see [7]. In higher dimensions, there are only a few theorems which, under special circumstances, guarantee that a reductive group action on $\mathbf{C}^n$ is conjugate to a linear action, see e.g. [3], [8]. A more detailed discussion of the Linearization Problem can be found in [1], [7], or Kraft's article in this volume. Not every unipotent subgroup of $\operatorname{Aut} \mathbf{C}^n$, $n \geq 3$, is conjugate to a subgroup of B_n. Bass [2] was the first to give an example in $\mathbf{C}^3$ and Popov has a simple technique for generating many such examples in any $\mathbf{C}^n$, see [10]. We shall review this construction in §2. We shall also examine some questions about free actions on $\mathbf{C}^n$ and their relationship to the Unipotent Subgroup Problem and to affine homogeneous spaces.

1. Vector fields.

When studying unipotent group actions there is naturally a special interest in the case $G \cong \mathbf{C}$. This is primarily due to the fact that a unipotent group has a composition series $G \triangleright G_1 \triangleright \ldots \triangleright G_k = 1$ with quotients $G_i/G_{i+1} \cong \mathbf{C}$. Such a series often allows one to generalize statements for $G \cong \mathbf{C}$ to arbitrary unipotent groups by induction on $\dim G$.

The simplest way to deal with an algebraic $\mathbf{C}$ action on $\mathbf{C}^n$ is to examine its associated vector field, which in our setting can be conveniently viewed as a locally nilpotent derivation of F_n. In fact, any locally nilpotent derivation determines (and is determined by) an algebraic $\mathbf{C}$ action in the following way. Recall that a derivation D of F_n is called locally nilpotent if there is a positive integer s such that $D^s x_i = 0$, $i = 1, \ldots, n$. Such a derivation generates a 1-parameter group, σ_t^*, of automorphisms of F_n:

$$\sigma_t^*(f) := \exp(tD)f = \sum_{m \geq 0} \frac{1}{m!} t^m D^m f \quad (t \in \mathbf{C},\ f \in F_n).$$

The sum, of course, is finite for any given f by Leibnitz's Rule and the fact that D is locally nilpotent. Conversely, given a 1-parameter subgroup, σ_t^*, of automorphisms of F_n, we obtain a locally nilpotent derivation:

$$f_i := \frac{\partial}{\partial t}\{\sigma_t^* x_i\}|_{t=0},$$

$$D := f_1 \frac{\partial}{\partial x_1} + \ldots + f_n \frac{\partial}{\partial x_n}.$$

It is easy to see that, under this association, the fixed points of the action of σ_t on $\mathbf{C}^n$ are precisely those points at which the functions $f_1, \ldots, f_n$ simultaneously vanish.

1.1. Examples. The following examples can be found in [**10**]. They will be referred to in the next section.

a) Define a 'linear' derivation Δ of $\mathbf{C}^n$ as follows. Choose a sequence of integers $p_1, \dots, p_s$ such that $1 \leq p_1 < \dots < p_s = n$, and let

$$\Delta x_i = \begin{cases} x_{i+1}, & \text{if } i \neq p_1, \dots, p_s \\ 0, & \text{if } i = p_1, \dots, p_s. \end{cases}$$

Then,

$$\sigma_t^* x_i = \exp(t\Delta)x_i = \sum_{k=0}^{p_j - i} \frac{1}{k!} t^k x_{i+k}$$

where p_j is the first integer in the sequence $\geq i$. By using a Jordan normal form, any unipotent 1-parameter subgroup of A_n can be obtained in this way. For future reference, we shall denote the above derivation Δ by ${}_n\Delta_{p_1,\dots,p_s}$.

b) Let D be a locally nilpotent derivation of F_n, and let $h \in F_n$ satisfy $Dh = 0$. Then h is invariant under $\exp(tD)$ and hD is again a locally nilpotent derivation of F_n generating a new 1-parameter subgroup of $\operatorname{Aut} \mathbf{C}^n$. It is not hard to see that at least one non-constant invariant always exists: Let W be the finite dimensional span of $\exp(tD)f$, $t \in \mathbf{C}$, for some fixed non-constant $f \in F_n$. Let W_0 be W modulo the invariant subspace of constant functions in W. By the Lie-Kolchin Theorem there is some non-zero vector $g \in W_0$ fixed by the linear action of $G = \{\exp(tD) \mid t \in \mathbf{C}\}$ on W_0. Viewed as a non-constant function on $\mathbf{C}^n$, g satisfies $\exp(tD)g = g + kt$. If $k = 0$ then g is an invariant; otherwise, the variety defined by $g = 0$ can be identified with the quotient space, Z, of G acting on $\mathbf{C}^n$ (see the discussion of equivariant slices in §3). Any algebraic function on Z pulled back to $\mathbf{C}^n$ by the quotient map is an invariant. □

2. Popov's construction.

Let us now summarize the construction of non-triangular actions of $\mathbf{C}$ on $\mathbf{C}^n$ given by Popov, [**10**]. Two lemmas are needed.

2.1. Lemma. *If σ_t, $t \in \mathbf{C}$, is a 1-parameter subgroup of* $\operatorname{Aut} \mathbf{C}^n$ *which is conjugate to a subgroup of B_n, then the variety of fixed points of σ_t, for any $t \neq 0$, is "cylindrical", that is, isomorphic to a product $\mathbf{C} \times Z$ for some affine variety Z.*

Proof. By assumption, there exists an automorphism γ of F_n such that

$$\gamma \sigma_t^* \gamma^{-1} x_i = c_i(t) x_i + h_i(t, x_{i+1}, \dots, x_n).$$

Now, c_i is a non-vanishing polynomial in t and $c_i(0) = 1$; hence $c_i(t) = 1$ for all t. Thus, the set of fixed points of $\gamma\sigma_t^*\gamma^{-1}$ is defined by the vanishing of $h_i(t, x_{i+1}, \ldots, x_n)$, $1 \le i \le n$, which altogether involve only the variables $x_2, \ldots, x_n$. Hence the fixed point set must be cylindrical. □

The second lemma describes the fixed point set of a 1-parameter subgroup generated by a derivation of the form $D = h\Delta$, where Δ is a derivation and h is an invariant; see the above examples.

2.2. Lemma. *Let Δ be a locally nilpotent derivation of F_n and let h be an invariant: $\Delta h = 0$. Define a new locally nilpotent derivation $D = h\Delta$ and let $\sigma_t^* = \exp(tD)$ be its associated 1-parameter group. Denote by $V(h)$ the hypersurface in $\mathbf{C}^n$ defined by $h = 0$. Then $V(h)$ is the union of certain irreducible components of the variety of fixed points of σ_t, $t \neq 0$, in $\mathbf{C}^n$.*

Proof. As noted above, the fixed points of σ_t are precisely where all of the coefficients of D vanish, and this certainly includes the variety $V(h)$. Now, $V(h)$ is a closed subvariety of pure dimension $n-1$ while the variety of fixed points has dimension $\le n-1$. Therefore, $V(h)$ is comprised entirely of certain irreducible components of the fixed point set. □

2.3. Corollary. *Let Δ be a locally nilpotent derivation of F_n with an invariant h such that the hypersurface $V(h)$ is not cylindrical. Then the 1-parameter unipotent group $\exp(th\Delta)$, $t \in \mathbf{C}$, is not conjugate to a subgroup of B_n.*

2.4. Example. If n is even, let Δ be the linear derivation $\Delta = {}_n\Delta_{n/2,n}$, see §1, Example *a*). This derivation leaves the non-degenerate quadratic form

$$h = \sum_{i=1}^{n/2} (-1)^i x_i x_{n+1-i}$$

invariant. Since $V(h)$ is clearly not cylindrical, the 1-parameter unipotent group $\exp(th\Delta)$ is not a 'triangularizable' subgroup of Aut $\mathbf{C}^n$. If n is odd, one may use $\Delta = {}_n\Delta_n$ and $h = \sum_{i=1}^{n} (-1)^i x_i x_{n+1-i}$. □

3. Free actions.

A group action is called 'free' if every isotropy group is trivial, i.e., if a group element fixes some point, then that element is the identity. Let G be an algebraic subgroup of Aut $\mathbf{C}^n$. If G acts freely on $\mathbf{C}^n$, then the group G must be unipotent. This is due to the fact that if G were not unipotent, then it would contain a subgroup isomorphic to $\mathbf{C}^*$ and this subgroup would necessarily have a fixed point in $\mathbf{C}^n$, see [4]. A special case which is easy to analyze via a Jordan normal form is when $G \cong \mathbf{C} \subset A_n$. If such a group G acts freely, then G is conjugate to a 1- parameter group of translations, and $\mathbf{C}^n$ is equivariantly isomorphic to $G \times \mathbf{C}^{n-1}$. Arbitrary free actions on $\mathbf{C}^n$ are not as well understood.

3.1. Problems. *Let G be an algebraic subgroup of* Aut $\mathbf{C}^n$ *acting freely on* $\mathbf{C}^n$.

1) *When does the quotient $\mathbf{C}^n/G$ exist as an algebraic variety?*
2) *If the quotient $\mathbf{C}^n/G$ exists, when is it an affine variety?*
3) *If the quotient $\mathbf{C}^n/G$ is an affine variety, when is it isomorphic to $\mathbf{C}^{n-1}$?*
4) *Is G conjugate to a subgroup of B_n? ('Are free actions triangular?')*
5) *Is G contained in a finite dimensional group which acts transitively on $\mathbf{C}^n$? ('Are free actions contained in transitive actions?')*

If the quotient $Y = \mathbf{C}^n/G$ does exist and is affine, and if the quotient map is smooth, then general arguments imply that $\mathbf{C}^n$ is equivariantly isomorphic to $G \times Y$ (see the proof of the next theorem, for example). Problem 3) is then equivalent to the notorious Cancellation Problem which asks whether an isomorphism $Y \times \mathbf{C} \cong \mathbf{C}^n$ implies that $Y \cong \mathbf{C}^{n-1}$. We have seen that Popov's construction gives an easy method for constructing non-triangular unipotent group actions on $\mathbf{C}^n$ which relies on the nature of the fixed point set. Perhaps the fixed point set being 'cylindrical' is the only obstruction for a $\mathbf{C}$ action to be triangular. In this light, problem 4) also seems reasonable. Finally, problem 5) brings us to questions about affine homogeneous spaces G/H. We shall return to this question in §4 where we shall see that if 5) is true then so is 4).

If one experiments with the above questions for the case of triangular actions on $\mathbf{C}^3$, it becomes clear how to overcome the difficulties by explicit computation.

3.2. Theorem. *Let G be a connected algebraic subgroup of B_3. If G acts freely on $\mathbf{C}^3$, then $\mathbf{C}^3$ is equivariantly isomorphic to $G \times \mathbf{C}^{3-\dim G}$.*

Proof. If $\dim G = 2$, then G is isomorphic to the direct product of two copies of the additive group $\mathbf{C}$. Assuming the theorem is true for one dimensional groups, the quotient of $\mathbf{C}^3$ by one of these $\mathbf{C}$ factors is $\mathbf{C}^2$. The remaining $\mathbf{C}$ factor still acts freely on this $\mathbf{C}^2$, and by the structure theorem for Aut $\mathbf{C}^2$, this latter action is conjugate to an action by translations. Consequently $\mathbf{C}^3/G \cong \mathbf{C}$, as claimed. Thus, the proof reduces to the case $\dim G = 1$. Let

$$D = f_1(x_2, x_3)\frac{\partial}{\partial x_1} + f_2(x_3)\frac{\partial}{\partial x_2} + f_3\frac{\partial}{\partial x_3}$$

be the derivation (vector field) associated to the action (since the action is triangular, the polynomial f_i depends only on $x_{i+1}, \dots, x_3$, $i = 1, 2, 3$. In particular, f_3 is constant). Notice that if $f_3 \neq 0$, we can replace x_2 by $x_2 - F_2/f_3$, where $F_2'(x_3) = f_2(x_3)$, so that D becomes the derivation $f_1\frac{\partial}{\partial x_1} + f_3\frac{\partial}{\partial x_3}$. Then, by changing x_1 in a similar way, D becomes $f_3\frac{\partial}{\partial x_3}$, i.e., the original action is conjugate to an action by translations and we are done. Thus, we may assume that $f_3 = 0$. Furthermore, since the action is free we know that the zero sets of f_1 and f_2 do not intersect. Applying the same reasoning as above, we may assume that neither f_1 nor f_2 is constant.

We construct a quotient map $\pi : \mathbf{C}^3 \to \mathbf{C}^2$ directly by writing down two invariant polynomials. Under our assumptions, x_3 is invariant. A second invariant is given by

$$h(x_1, x_2, x_3) := \int_0^{x_2} f_1(u, x_3)\, du \quad - \quad x_1 f_2(x_3).$$

We now define $\pi(x_1, x_2, x_3) = (h(x_1, x_2, x_3), x_3)$. Since π is defined by invariant functions, a π-fiber contains the orbit of any of its points. Moreover, the dimension of any fiber is clearly 1, so that each fiber of π consists of a disjoint union of orbits. Direct calculations show, however, that these fibers consist of only one orbit each, and that π is surjective [**16**]. Finally, π is smooth because its Jacobian is $\begin{bmatrix} -f_2 & f_1 & * \\ 0 & 0 & 1 \end{bmatrix}$ which always has rank two, since f_1, f_2 never simultaneously vanish. Therefore, the implicit function theorem guarantees the existence of local analytic sections to the map π. Therefore, $\pi : \mathbf{C}^3 \to \mathbf{C}^2$ is a principal $\mathbf{C}$-bundle which is locally trivial in the étale topology, and hence globally trival, see [**12**]. □

3.3. Equivariant Slices. For a free action of a group G on a space X, one obtains an equivariant trivialization $X \cong (X/G) \times G$ precisely when there is an equivariant map $s : X \to G$. Such a map defines a 'slice', $S := s^{-1}(1) \subset X$, for the action of G on X, that is, S intersects every G-orbit in X in exactly one point. For this reason we shall call the map itself an 'equivariant slice'. One can identify S with the orbit space, X/G: For each $x \in X$ there is a unique $g \in G$ such that $s(g \cdot x) = 1$, namely $g = s(x)^{-1}$. Then the quotient map, $\pi : X \to S \cong X/G$, is $\pi(x) = s(x)^{-1} \cdot x$. A trivialization is defined by sending $x \in X$ to $(\pi(x), s(x)) \in (X/G) \times G$. Conversely, given an equivariant trivialization, $X \to (X/G) \times G$, an equivariant slice is obtained by composing this trivialization with projection onto the second factor.

When dealing with free actions of $G \cong \mathbf{C}$ on $\mathbf{C}^n$, an equivariant slice is simply a function, $s : \mathbf{C}^n \to \mathbf{C}$, such that $s(\sigma_t(x)) = s(x) + t$. Let D be the locally nilpotent derivation associated to the action. Since

$$Df = \frac{\partial}{\partial t}\{\sigma_t^* f\}|_{t=0},$$

we see that an equivariant slice s must satisfy $Ds = 1$. Moreover, a solution to the equation $Ds = 1$ clearly defines an equivariant slice.

3.4. Example. Perhaps the simplest action of $G = \mathbf{C}$ on $\mathbf{C}^3$ which is not obviously conjugate to a group of translations is the one induced by the following derivation:

$$D := (1 - x_2x_3)\frac{\partial}{\partial x_1} + x_3^2\frac{\partial}{\partial x_2}.$$

Since D has triangular form, it is automatically locally nilpotent. The corresponding 1-parameter group of automorphisms, σ_t, of $\mathbf{C}^3$ is given by

$$\sigma_t(x_1, x_2, x_3) = (x_1 + (1 - x_2x_3)t - \frac{1}{2}x_3^2t^2, x_2 + x_3t^2, x_3).$$

An equivariant slice s for this action is

$$s(x_1, x_2, x_3) = -\frac{1}{2}\{x_1(3 - (1 - x_2x_3)^2 - x_2x_3^3) + x_2^3(1 - \frac{1}{4}x_2x_3)\},$$

which can be found by explicitly solving

$$(1 - x_2x_3)\frac{\partial s}{\partial x_1} + x_3^2\frac{\partial s}{\partial x_2} = 1$$

for the coefficients in a polynomial expression for s. □

The above theorem also holds for quasi-algebraic actions of $G = \mathbf{C}$ on $\mathbf{C}^3$, see **[16]**. These are actions given by an analytic map $\mu : G \times \mathbf{C}^3 \to \mathbf{C}^3$ such that for each fixed $g \in G$, $\mu(g, \ ) : \mathbf{C}^3 \to \mathbf{C}^3$ is an algebraic automorphism.

At the time I presented some of the questions above at the conference at Rutgers, no examples were known of free $\mathbf{C}$ actions on $\mathbf{C}^n$ which did not have a quotient isomorphic to $\mathbf{C}^{n-1}$. Since that time I have heard from Martha Smith, University of Texas at Austin, and Jörg Winkelmann, Ruhr-Universität Bochum, who have constructed some interesting examples of badly behaved free $\mathbf{C}$ actions on $\mathbf{C}^n$ for $n \geq 4$. The following is a modified version of Smith's example, with her algebraic proof of the non-existence of a quotient replaced by a simple geometric argument. Winkelmann also has found such an example, see [**17**].

3.5. Example. Let D be the derivation

$$D = (1+z^2)D_x + zD_y + wD_z.$$

Computing the $\mathbf{C}$ action that we get from D we find that for $t \in \mathbf{C}$

$$t \cdot (x,y,z,w) = (x + (1+z^2)t + zwt^2 + \frac{1}{3}w^2t^3, y + zt + \frac{1}{2}wt^2, z + wt, w).$$

Now let $p_\epsilon = (0,0,i\sqrt{3},\epsilon)$ and $q_\epsilon = (0,0,-i\sqrt{3},\epsilon)$. Both p_ϵ and q_ϵ li e in the same orbit if $\epsilon \neq 0$ since

$$t \cdot (0, \frac{3}{2\epsilon}, 0, \epsilon) = \begin{cases} p_\epsilon, \text{if } t = i\sqrt{3}/\epsilon \\ q_\epsilon, \text{if } t = -i\sqrt{3}/\epsilon. \end{cases}$$

However, as $\epsilon \to 0$, $p_\epsilon \to p_0$ and $q_\epsilon \to q_0$. Thus the single orbit $\mathbf{C} \cdot p_\epsilon$ approaches the two distinct orbits $\mathbf{C} \cdot p_0$ and $\mathbf{C} \cdot q_0$ simultaneously as $\epsilon \to 0$. This shows that the topological orbit space cannot be Hausdorff. □

4. Homogeneous spaces.

Let G be a connected linear algebraic group, and let H be an algebraic subgroup. In the theory of homogeneous spaces it is useful to have group theoretic criteria which guarantee certain geometric properties of the coset space G/H. For example, G/H is a projective variety if and only if H contains a maximal solvable subgroup of G. Another example is Matsushima's Theorem [**9**]: If G is a reductive group, then G/H is an affine variety if and only if H is also reductive. This theorem is true in any characteristic, see [**11**]. If G is an arbitrary linear algebraic group, little is known about group theoretic conditions on H which imply that G/H is an affine variety. The characterizations of general homogeneous affine varieties so far are of a cohomological nature, see [**5**], [**6**], [**14**]. In this section we shall discuss an analogue of Matsushima's criterion for arbitrary linear algebraic groups and its relationship to free triangular actions on affine space $\mathbf{C}^n$.

4.1. Definition. *A subgroup H of a linear algebraic group G satisfies Matsushima's criterion if the unipotent radical of the connected component of H has only trivial intersection with any reductive subgroup of G.*

Obviously, if G is a reductive subgroup, then an algebraic subgroup satisfying Matsushima's criterion has to be reductive. The connection to homogeneous affine varieties is the following.

4.2. Lemma. *If G/H is an affine variety, then H satisfies Matsushima's criterion.*

Proof. Since the map $G/H^0 \to G/H$ is finite, G/H is affine if and only if G/H^0 is affine, so we may as well assume that H is connected. Write G as a semidirect product, $G = M \cdot U$, where M is a maximal reductive subgroup, and U is the unipotent radical of G. Let $H = L \cdot V$ be a similar decomposition for H. By the conjugacy of maximal reductive subgroups, we may assume that $L \subset M$. However, we do not know where the subgroup V is located with respect to U; in particular V is not necessarily a subgroup of U. The natural map of cosets $G/V \to G/H$ is a principal L-bundle. Since L is reductive, it is well-known that the total space, G/V, is affine if and only if the base, G/H, is affine, see e.g. [**6**], Lemma 4.1.

As with any algebraic action, the orbits of M in G/V which have minimal dimension (among the orbits of M) are closed. Let $M.x$ be such a closed orbit which must be affine since G/V is. Matsushima's theorem implies that the isotropy group M_x is reductive. Now $x = g_0V$, for some $g_0 \in G$, and $M_x = M \cap g_0Vg_0^{-1}$ is a reductive subgroup of the unipotent group $g_0Vg_0^{-1}$. This is impossible unless $M_x = 1$. We conclude that this orbit, $M.x$, of *minimal dimension* actually has the *maximum* dimension possible, $\dim M$. In particular, for any $g \in G$, $y = gV \in G/V$: $\dim M \geq \dim M.y \geq \dim M.x = \dim M$. It follows that for any $g \in G$, $M \cap gVg^{-1} = 1$. By the conjugacy of maximal reductive subgroups, this implies that V has only trivial intersection with any reductive subgroup of G. □

4.3. Remarks. If G is a complex Lie group acting effectively and transitively on a Stein manifold X, then the isotropy subgroups must also satisfy Matsushima's criterion. In fact, for isotropy subgroups H having finitely many connected components, the theory of Stein quotients G/H closely parallels the theory of affine quotients, see [**14**].

The proof of the previous lemma highlights an important geometric property of a subgroup H of G satisfying Matsushima's criterion: The action of M on the coset space G/V is *free* (we retain the previous notation). Another way to express this is to say that the unipotent group V acts freely, from the right, on $M\backslash G \cong U \cong \mathbf{C}^n$. Thus, the converse to the above lemma brings us back to the study of free unipotent actions on affine space. In this situation there is an additional property: because of the unipotent group structure on U, it is not hard to show that the action of V on U is conjugate to a triangular action, see [**16**]. Problem 3.1.5 asks whether every free action of a (necessarily unipotent group) V on $\mathbf{C}^n$ can be imbedded in a transitive action of an algebraic group G. If this were true, then the isotropy subgroup H would have to contain a maximal reductive subgroup M of G and the unipotent radical would act transitively on $\mathbf{C}^n \cong G/H \cong U/U \cap H$. Therefore, V would satisfy Matsushima's criterion in G. Furthermore, if $U \cap H = 1$, then the action of V on $\mathbf{C}^n \cong U$ would be triangular (Problem 3.1.4). The two areas are connected in many such ways, and results in one area translate directly into results in the other.

In the above setting, if $\dim V = 1$ or if $\dim U \leq 3$ then the quotient U/V exists and is isomorphic to affine space, see [**6**], and [**16**], respectively. As attractive as this may seem, it is not the general pattern. Winkelmann [**17**] has recently constructed an example of a two-dimensional subgroup $V \subset G = A_6$ which acts freely on the unipotent radical $U \cong \mathbf{C}^6$ of G such that the quotient U/V exists, is a quasi- affine variety homeomorphic to $\mathbf{C}^4$, but is *not* an affine variety. Therefore, Matsushima's criterion alone is not sufficient, in general, to decide whether a homogeneous space is affine. Winkelmann's example is the following:

4.4. Example. Let V be the two dimensional subgroup of $G = A_6$ defined by the following affine linear transformations of $z = (z_1, \dots, z_6) \in \mathbf{C}^6$:

$$\phi_{s,t}(z) = (z_1+s+tz_2+\frac{1}{2}t^2(1+z_6), z_2+t(1+z_6), z_3+tz_5, z_4+tz_6+sz_5, z_5, z_6)$$

where $s, t \in \mathbf{C}$. Since $\phi_{s,t}(z) = z \Rightarrow s = t = 0$, the action of V on $\mathbf{C}^6$ is free. The algebra of invariant polynomials for the one-dimensional subgroup $V_1 = \{\phi_{s,0} \,|\, s \in \mathbf{C}\} \subset V$ is obviously generated by $z_2, z_3, z_4 - z_1z_5, z_5, z_6$. Since V is abelian, the action of the one-dimensional subgroup $V_2 = \{\phi_{0,t} \,|\, t \in \mathbf{C}\} \subset V$ passes to the quotient $\mathbf{C}^6/V_1 \cong \mathbf{C}^5$. Making the substitutions $u_5 = z_6 + 1$, $u_4 = z_5$, $u_3 = z_2$, $u_2 = z_3$, and $u_1 = z_4 - z_1z_5$, the resulting action on $u = (u_1, \dots, u_5) \in \mathbf{C}^5$ is given by:

$$\phi_t(u) = (u_1+t(u_5-1)-tu_3u_4-\frac{1}{2}t^2u_4u_5, u_2+tu_4, u_3+tu_5, u_4, u_5),\ t \in \mathbf{C}.$$

This can be simplified by substituting $w_1 = -u_1 + u_3 - u_3u_2/2$ and $w_k = \lambda u_k$, $2 \le k \le 5$, where $\lambda = \sqrt{-1/2}$, to obtain:

$$\phi_t(w) = (w_1 + t(1 + d(w)), w_2 + tw_4, w_3 + tw_5, w_4, w_5)$$

where $d(w) = w_3w_4 - w_2w_5$. Invariants of this action are $\xi_1 = d(w)$, $\xi_2 = w_2(1 + d(w)) - w_1w_4$, $\xi_3 = w_3(1 + d(w)) - w_1w_5$, $\xi_4 = w_4$, and, $\xi_5 = w_5$. These functions define a map $\xi : \mathbf{C}^5 \rightarrow \mathbf{C}^5$ whose image lies in the non-singular quadric $Q = \{x \in \mathbf{C}^5 \mid x_1(x_1+1) - x_2x_5 + x_3x_4 = 0\}$. It is not hard to show that the fibers of this map are precisely the orbits of V_2 in $\mathbf{C}^5$ and that the map ξ has maximal rank everywhere. Direct calculation reveals the surprising fact that the image, $\xi(\mathbf{C}^5)$, is actually $Q \backslash E$, where $E = \{x \in Q \mid x_1 = -1, x_4 = x_5 = 0\}$. Since E has codimension 2 in Q, the quotient space $\mathbf{C}^6/V \cong Q\backslash E$ is not affine. □

References

[1] Bass, H.: Algebraic group actions on affine space. *Contemporary Math.* **43,** 1–23 (1985)

[2] Bass, H.: A non-triangular action of G_a on A^3. *J. Pure Appl. Alg.* **33,** 1–5 (1984)

[3] Bass, H., Haboush, W.: Linearizing reductive group actions. *Trans. Amer. Math. Soc.* **292,** 463–482 (1985)

[4] Białynicki-Birula, A.: Remarks on the action of an algebraic torus on k^n, I. *Bull. Acad. Pol. Sci.* **14,** 177– 181 (1966)

[5] Borel, A.: On affine algebraic homogeneous spaces. *Arch. Math.* **45,** 74–78 (1985)

[6] Cline, E., Parshall B., Scott L.: Induced modules and affine quotients. *Math. Ann.* **230,** 1–14 (1974)

[7] Kraft, H.: Algebraic group actions on affine spaces. In: Geometry of Today (Roma 1984). Boston, Basel, Stuttgart: Birkhäuser 1985.

[8] Kraft, H., Popov, V.: Semisimple group actions on the three dimensional affine space are linear. *Comment. Math. Helv.* **60,** 466–479 (1985)

[9] Matsushima, Y.: Espaces homogenènes de Stein des groupes de Lie complexes. *Nagoya Math. J.* **18,** 153–164 (1961)

[10] Popov, V.: On actions of G_a on A^n. In: Algebraic Groups (Utrecht 1986), Lecture Notes in Math. vol. 1271. Berlin, Heidelberg, New York: Springer 1987.

[11] Richardson, R.: Affine coset spaces of affine algebraic groups. *Bull. London. Math. Soc.* **91,** 38–41 (1977)

[**12**] Serre, J.P.: Espaces fibrés algébriques. In: Anneaux de Chow et Applications. Séminaire Chevelley, Paris 1958.

[**13**] Snow, D.: Reductive group actions on Stein spaces. *Math. Ann.* **259,** 79–97 (1982)

[**14**] Snow, D.: Stein quotients of connected complex Lie groups. *Manuscr. Math.* **50,** 185–214 (1985)

[**15**] Snow, D.: Invariants of holomorphic affine flows. *Arch. Math.* **49,** 440–449 (1987)

[**16**] Snow, D.: Triangular actions on C^3. *Manuscr. Math.* **76,** 1–10 (1988)

[**17**] Winkelmann, J.: On Stein homogeneous manifolds and free holomorphic C actions on C^n. *preprint*

Dennis M. Snow
Department of Mathematics
University of Notre Dame
Notre Dame, IN 46556
USA

ALGEBRAIC CHARACTERIZATION OF THE AFFINE PLANE AND THE AFFINE 3–SPACE

Toru Sugie

§ 0. Introduction

In affine algebraic geometry, it is very important to find a good characterization of the affine n-space $\mathbf{A}^n$ over $\mathbf{C}$ as an algebraic variety. For example, we have the following characterization of the affine plane $\mathbf{A}^2$ over $\mathbf{C}$.

Theorem 1 [6]. *Let $X = \mathrm{Spec}A$ be an affine surface over $\mathbf{C}$. Suppose that the following three conditions are satisfied:*

A-1) $A^* = \mathbf{C}^*$, *where A^* is the group of invertible elements of A,*
A-2) *A is factorial,*
A-3) *X contains a cylinderlike open set $U \simeq U_0 \times \mathbf{A}^1$ where U_0 is an affine curve.*

Then X is isomorphic to the affine plane $\mathbf{A}^2$.

In many applications we find it difficult to check the third condition of the theorem. But when X is nonsingular, by the following results in [1] and [7], we can replace the third condition by the following condition A-4).

Theorem 2. *Let $X = \mathrm{Spec}A$ be a smooth affine surface over $\mathbf{C}$ as above. Embed X into a smooth projective surface, say V, as a Zariski open set so that the boundary divisor $D := V - X$ consists of smooth curves with only normal crossings as its singularities. Let K_V be the canonical divisor of V. Then the condition A-3) is equivalent to the following condition A-4).*

A-4) $h^0(V, O(m(K_V + D))) = 0$ *for* $\forall m \geq 1$.

Remark: The condition A-4) does not depend on the choice of an open embedding X into a smooth projective surface V. The condition A-4) is equivalent to say that the logarithmic Kodaira dimension of X is $-\infty$.

Using Theorem 2, we can prove the following:

Cancellation Theorem [1],[7]. *Let X be an affine variety defined over $\mathbf{C}$. Assume that $X \times \mathbf{A}^n \simeq \mathbf{A}^{n+2}$. Then X is isomorphic to an affine plane $\mathbf{A}^2$ over $\mathbf{C}$.*

In this note the notation $\simeq$ signifies the isomorphism as algebraic varieties. For other applications of Theorem 2, see [9].

The next problem is how to characterize the affine 3-space $\mathbf{A}^3$. There are several attempts to find such a characterization. Among them we have the following theorem by Miyanishi.

Theorem 3 [13]. *Let A be an affine domain over $\mathbf{C}$ of dimension 3 and let $X = \mathrm{Spec}(A)$. Then X is isomorphic to $\mathbf{A}^3$ if and only if the following conditions are satisfied.*

B-1) *$A^* = \mathbf{C}^*$, where A^* is the group of invertible elements of A,*
B-2) *A is factorial,*
B-3) *X contains a nonempty Zariski open set U such that U is isomorphic to $U_0 \times \mathbf{A}^2$, where U_0 is an affine curve, and that each irreducible component of $X - U$ has the coordinate ring which is factorial,*
B-4) *X has the topological Euler number 1.*

To apply to the cancellation theorem for the affine 3-space, we don't have a sufficient characterization. In this direction we have only a partial answer. To state a result, we need the following definitions.

Given a morphism $f : X \to Y$ of algebraic $\mathbf{C}$-schemes, we say that f is geometrically irreducible in codimension one over Y if, for any irreducible subvariety T of codimension one of X, the function field extension $\mathbf{C}(T)$ over $\mathbf{C}(\overline{f(T)})$ is a regular extension, where $\overline{f(T)}$ is the closure of $f(T)$ in Y.

Let A be an affine domain over $\mathbf{C}$ of dimension n and let $X = \mathrm{Spec}(A)$. We say that X is affine 1-ruled if X contains a nonempty Zariski open set Z such that Z is isomorphic to $Z_0 \times \mathbf{A}^1$. If X is affine 1-ruled, G_a acts on X(see Lemma 1). Set $Y = \mathrm{Spec}(A^{G_a})$. Then we obtain a morphism $f : X \to Y$ from the inclusion $A^{G_a} \hookrightarrow A$.

Under the above notations, we have the following:

Theorem 4. *Let A be an affine domain over $\mathbf{C}$ of dimension three and let $X = \mathrm{Spec}(A)$. Assume that following three conditions hold.*

C-1) *$X \times \mathbf{A}^n \simeq \mathbf{A}^{n+3}$,*
C-2) *X is affine 1-ruled,*

C-3) $f : X \to Y$ *(defined as above) is geometrically irreducible in codimension one over* Y.

Then X *is isomorphic to* $\mathbf{A}^3$.

We remark that in the case of dimension three, the conditions B-1), B-2) and C-2) do not imply $X \simeq \mathbf{A}^3$. See a remark in §4.

In this report in §1 we give a sketch of the proof of Theorem 1. In §2 we indicate how to prove the equivalence of two conditions A-3) and A-4) and explain Fujita's theory of pseudo-effective divisors. In §3 we remark the relation between the above algebraic characterization of $\mathbf{A}^2$ and the topological one. Finally in §4 we give an outline of the proof of Theorem 4. Theorem 4 is a joint result with M. Miyanishi.

§ 1. Algebraic characterization of $\mathbf{A}^2$

We give a proof of Theorem 1. First we prove the following:

Lemma 1. *Let* A *be an affine domain over* $\mathbf{C}$ *and put* $X = \mathrm{Spec}A$. *We assume that* A *is factorial. Then the following three conditions are equivalent.*

(1) X *is affine 1-ruled,*
(2) *There exists a nontrivial locally nilpotent* $\mathbf{C}$*-derivation* Δ *of* A.
(3) G_a *acts on* X.

PROOF: If G_a acts on X, then it is easy to see that X is affine 1-ruled.

(1) $\longrightarrow$ (2) Assume that X is affine 1-ruled. Then X contains an affine open set Z such that Z is isomorphic to $Z_0 \times \mathbf{A}^1$. Since Z is affine, $X - Z$ is pure of codimension one and is the support of a divisor on X. But X is factorial, so $X - Z = \mathrm{Supp}(s)$ for some principal divisor (s) with $s \in A$. Therefore putting $Z_0 = \mathrm{Spec}(B)$ with a $\mathbf{C}$-algebra B, we have

$$A_s = A[s^{-1}] = B[t]$$

where t is transcendental over B. Since s is a unit in A_s and a unit in $B[t]$ is contained in B, so $s \in B$.

Now we put $D = \frac{\partial}{\partial t}$. It is a derivation of $B[t]$ which is trivial on B. Since A is finitely generated over $\mathbf{C}$, we can write $A = \mathbf{C}[a_1, a_2, \cdots, a_r]$ with $a_i \in A$. Then there exists an integer l such that $s^l Da_i \in A$ for $1 \leq i \leq r$. Put $\Delta = s^l D$. We choose l such that $\Delta(A) \subseteq A$. Then since $D(s) = 0$, Δ defines a locally nilpotent derivation of A.

(2) $\longrightarrow$ (3) Assume that there exists a nontrivial locally nilpotent $\mathbf{C}$-derivation Δ of A. Define a $\mathbf{C}$-algebra homomorphism $\sigma : A \to A[u]$ by

$$\sigma(a) = a + \Delta(a)u + \cdots + \frac{1}{n!}\Delta^n(a) + \cdots$$

where $a \in A$ and u is transcendental over A. Since Δ is locally nilpotent, the right hand side of the above equation is a polynomial. It is easy to check that the morphism $X \times \mathbf{A}^1 \to X$ induced by σ defines an action of G_a on X.

Proof of Theorem 1: Now we assume $\dim_{\mathbf{C}} A = 2$ and assume that the three conditions in Theorem 1 hold. We will give a proof of Theorem 1 when X is smooth. Then by Lemma 1, G_a acts on $X = \mathrm{Spec}A$. Put $A^{G_a} = A_0$. By Zariski's lemma, A_0 is finitely generated over $\mathbf{C}$. Put $C = \mathrm{Spec}A_0$. Put $\pi : U = U_0 \times \mathbf{A}^1 \to U_0$ be a natural projection and let π' be the morphism associated with the natural inclusion $A_0 \hookrightarrow A$. Then we have the following commutative diagram (note that the proof of Lemma 1 shows G_a acts on $U_0 \times \mathbf{A}^1$ along $\mathbf{A}^1$):

$$\begin{array}{ccc} U \simeq U_0 \times \mathbf{A}^1 & \longrightarrow & X \\ \downarrow{\scriptstyle \pi} & & \downarrow{\scriptstyle \pi'} \\ U_0 & \longrightarrow & C \end{array}$$

The first two assumptions imply that C is a smooth affine curve and $\mathrm{Pic}(C) = 0$. This implies C is rational. Since $A^* = \mathbf{C}^*$ and $\mathrm{Pic}(C) = 0$, C is isomorphic to $\mathbf{A}^1$. Now examining contributions to $\mathrm{Pic}(X)$ from singular fibers of π', it is easy to see that every fiber of π' is smooth and isomorphic to $\mathbf{A}^1$. See [8] for example. Thus, X is an $\mathbf{A}^1$-bundle over $\mathbf{A}^1$, which has to be trivial. We conclude that $X \simeq \mathbf{A}^2$.

§ 2. Cancellation Theorem for the Affine Plane $\mathbf{A}^2$

In this section we give an outline of the proof of Theorem 2 and the Cancellation Theorem for $\mathbf{A}^2$. First we recall the notion of the logarithmic Kodaira dimension.

Let X be a nonsingular algebraic variety defined over $\mathbf{C}$. Then there exists a nonsingular projective variety V such that X is a dense open subset, $V - X$ has pure codimension one in V and $V - X$ has only normal crossings as its singularities. Let D be the reduced effective divisor with

$\mathrm{Supp}(D) = V - X$. Let K_V be the canonical divisor of V. Define the Kodaira dimension $\kappa(X)$ of X as follows:

$$\kappa(X) = \begin{cases} -\infty, & \text{if } |n(D+K_V)| = \emptyset \text{ for all } n > 0 \\ \sup_{n>0} \dim \Phi_{|n(D+K_V)|}(V), & \text{otherwise} \end{cases}$$

$\kappa(X)$ does not depend on the choice of an open embedding of X into a smooth projective variety V. $\kappa(X)$ takes one of the values $-\infty$, 0, 1, 2. See [4] for an explicit definition and various properties of $\kappa(X)$. Here we cite the following:

Lemma 2. *Let $f : X \to Y$ be a dominant morphism of nonsingular algebraic varieties of dimension n. Then we have*

$$\kappa(X) \geq \kappa(Y).$$

Now we show how Theorem 2 implies Cancellation Theorem. Let X be an affine variety defined over $\mathbf{C}$ and assume that $X \times \mathbf{A}^n \simeq \mathbf{A}^{n+2}$. Then clearly X is a nonsingular affine surface and satisfies the conditions A-1) and A-2) of Theorem 1. Take an affine plane $\mathbf{A}^2$ in $\mathbf{A}^{n+2}$ such that the morphism

$$\tau : \mathbf{A}^2 \xrightarrow{i} \mathbf{A}^{n+2} \simeq X \times \mathbf{A}^n \xrightarrow{\pi} X$$

is dominant, where i is the inculusion and π is the natural projection. Since $\kappa(\mathbf{A}^2) = -\infty$, Lemma 2 implies that $\kappa(X) = -\infty$. This is the condition A-4) of Theorem 2. Thus to prove the Cancellation Theorem, it is sufficient to prove the equivalence of A-3) and A-4).

We try to prove the equivalence by extending the following well-known Enriques characterization of rational or ruled surfaces.

Theorem 5 [15]. *Let F be a nonsingular projective surface without exceptional curves of the first kind. Then the following four conditions are equivalent:*

E-1) *F is isomorphic to the projective plane or a $\mathbf{P}^1$-bundle over a complete nonsingular curve,*

E-2) *there exists an irreducible curve C on F such that $(C \cdot K_F) < 0$,*

E-3) *for any divisor A on F, $|A + nK_F| = \emptyset$, for every sufficiently large integer n,*

E-4) *$|nK_F| = \emptyset$ for every positive integer n, i.e., $\kappa(F) = -\infty$.*

We extend the above theorem to the case of noncomplete nonsingular algebraic surfaces. Let X be a nonsingular affine surface and define V, D and K_V as in Theorem 2. We consider the following four conditions:

F-1) X is affine 1-ruled,

F-2) there exists an irreducible curve C on V such that C is not contained in Supp(D) and $(C \cdot D + K_V) < 0$,

F-3) for any divisor A on V, $|A + n(D + K_V)| = \emptyset$ for every sufficiently large integer n,

F-4) $|n(D + K_V)| = \emptyset$ for every positive integer n, i.e., $\kappa(X) = -\infty$.

The implication F-1) → F-2) → F-3) → F-4) is easy. In [7] we proved the equivalence of the three conditions F-1), F-2) and F-3). In [1] Fujita proved the implication F-4) → F-3). For this we need his theory of pseudo-effective divisors.

We define a $\mathbf{Q}$-divisor L on V to be a finite sum of irreducible curves with rational coefficients: $L = \Sigma x_i C_i, x_i \in \mathbf{Q}$, where C_i are irreducible curves on V.

We say L is effective if $x_i \geq 0$ for all i. L is arithmetically effective if $(L \cdot C) \geq 0$ for all irreducible curves C on V. L is pseudo-effective if $(D \cdot H) \geq 0$ for every arithmetically effective $\mathbf{Q}$-divisor on H. Clearly an effective $\mathbf{Q}$-divisor is pseudo-effective. Then we have the following theorem due to Fujita.

Theorem 6 [1]. *Let L be a pseudo-effective $\mathbf{Q}$-divisor on V. Then L is numerically equivalent to the sum of $\mathbf{Q}$ divisors P and N, which satisfy*

(1) *P is arithmetically effective,*

(2) *N is effective,*

(3) $(P \cdot N) = 0$,

(4) *the intersection matrix of the support of N is negative definite.*

We call the above decomposition the Zariski decomposition of L. Using this decomposition Fujita proved the following:

Lemma 3 [1]. *Let L and M be divisors on V. Assume there exists an infinite sequence of integers $\{n_1, n_2, \cdots\}$ such that $\lim_{j\to\infty} n_j = \infty$ and $\dim_{\mathbf{C}} H^0(V, L + n_j M) \geq 1$ for all j. Then M is pseudo-effective.*

Lemma 4 [1]. *Let V and D be as above. Above all, D is a reduced effective divisor on V with only simple normal crossings as singularities. Assume that $D + K_V$ is pseudo-effective. Then there exists an integer m such that $|n(D + K_V)| \neq \emptyset$.*

The above two lemmas show the implication F-4) → F-3). This proves Theorem 2 and also finishes the proof of the Cancellation Theorem.

§ 3. Topological characterization of A^2

We remark the relation between the algebraic characterization of $\mathbf{C}^2$ and

Ramanujam's.

Ramanujam's characterization of $\mathbf{C}^2$ [16]. *Let X be a nonsingular algebraic surface defined over $\mathbf{C}$. Assume that X is contractible and the fundamental group at infinity of X is trivial, i.e., $\pi_1^\infty(X) = (e)$. Then X is isomorphic to $\mathbf{C}^2$.*

In [3] Gurjar-Miyanishi proved the following theorem:

Theorem 7 [3]. *Let X be a nonsingular affine surface defined over $\mathbf{C}$. Assume that the fundamental group at infinity of X is a finite group. Then $\kappa(X) = -\infty$.*

When X is a contractible surface, it is easy to see that X is an affine surface and X satisfies the conditions A-1) and A-2)(See [2]). Therefore by Theorem 1 and Theorem 2 we deduce Ramanujam's Characterization from Theorem 7. Moreover in Ramanujam's Characterization we can replace the condition $\pi_1^\infty(X) = (e)$ by $\pi_1^\infty(X)$ is a finite group. See also [4].

§ 4. Cancellation of $\mathbf{A}^3$

In this section we give a proof of Theorem 4. We need several lemmas. First we cite the following:

Theorem 8 [12]. *Let G be a connected linear algebraic group defined over $\mathbf{C}$ and let G act on an affine n-space $\mathbf{C}^n$ algebraically. We assume that following two conditions hold:*

(1) *G has no nontrivial multiplicative characters;*
(2) *the affine quotient variety $\mathbf{C}/G = \mathrm{spec}\mathbf{C}[X_1, \cdots, X_n]^G$ is a surface.*

Then $\mathbf{C}/G$ is isomorphic to an affine plane $\mathbf{A}^2$ over $\mathbf{C}$.

Note that $A = \mathbf{C}[X_1, \cdots, X_n]^G$ is finitely generared by Zariski lemma. We will give a proof of this theorem depending on [12]. We need some results for the proof of Theorem 8.

Theorem 9 [12]. *Let X be a normal affine surface defined over $\mathbf{C}$ such that $\Gamma(X, O_X)$ is factorial. Suppose that there exists a dominant morphism $f : \mathbf{C}^n \to X$ such that $f^{-1}(P) = \emptyset$ or $\dim f^{-1}(P) \leq n-2$ for every singular point of X. Then either X is isomorphic to $\mathbf{C}^2$ or X is isomorphic to the hypersurface $x^2 + y^3 + z^5 = 0$ in $\mathbf{C}^3$.*

To prove this Theorem 9 we first need several notations and results from [12].

A nonsingular algebraic surface T is called a $\mathbf{C}^*$-fiber space if there exists a surjective morphism $f : T \to C$ from T to a nonsingular algebraic curve C such that general fibers of f are isomorphic to $\mathbf{C}^* = \mathbf{C} - 0$. A fiber Δ of f is called a singular fiber if either Δ is reducible or Δ has the form $\Delta = \mu\Gamma$ with $\mu \geq 2$ and Γ irreducible: in the latter case, Δ is a multiple fiber with multiplicity μ. Given a $\mathbf{C}^*$-fiber space $f : T \to C$ with C complete, there exists a nonsingular projective surface V and a surjective morphism $\pi : V \to C$ such that

(1) T is a Zariski open set of V and $V - T$ is a reduced effective divisor with at most simple normal crossings as its singularities;

(2) general fibers of π are isomorphic to $\mathbf{P}^1_{\mathbf{C}}$ and the restriction of π onto T coincides with the given morphism.

We call $\pi : V \to C$ a normal completion of $f : T \to C$. A $\mathbf{C}^*$-fiber space $f : T \to C$ is called Platonic if the following conditions are satisfied.

(1) C is isomorphic to $\mathbf{P}^1_{\mathbf{C}}$;

(2) f has exactly three singular fibers $\Delta_i = \mu_i\Gamma$ $(1 \leq i \leq 3; \mu_1 \leq \mu_2 \leq \mu_3)$ with $\Gamma_i \simeq \mathbf{C}^*$, where $\{\mu_1, \mu_2, \mu_3\} = \{2, 2, n\} (n \geq 2)$, $\{2, 3, 3\}, \{2, 3, 4\}$, or $\{2, 3, 5\}$;

(3) $f : T \to C$ has a normal completion $\pi : V \to C$ such that

(a) there exist two cross-sections S_0 and S_1 of π such that $S_0, S_1 \subset V - T$ and $S_0 \cap S_1 = \emptyset$, and other irreducible components of $V - T$ are contained in the fibers of π;

(b) every fiber of π has a linear chain as its dual graph.

Note that if $\pi : V \to C$ is a normal completion of $f : T \to C$ as above, then the boundary $V - T$ consists of two connected components R_0 and R_1 and one of them, say R_0 contains S_0 and R_1 contains S_1.

The structure of a Platonic $\mathbf{C}^*$-fiber space is given in [**12**]. Before stating the result we recall that an element g of $GL(2, \mathbf{C})$ is called a pseudo-reflection if $\operatorname{rank}(g - I) \leq 1$ and that a finite subgroup G of $GL(2, \mathbf{C})$ is called small if G contains no pseudo-reflections.

Lemma 5 [**12**]. *(1) Let T be a Platonic $\mathbf{C}^*$-fiber space. Then $\mathbf{C}^2 - 0$ is the universal covering space of T, $G := \pi_1(T)$ is a small finite subgroup of $GL(2, \mathbf{C})$ which is not a cyclic group and T is isomorphic to the quotient variety $\mathbf{C}^2/G$ with the unique singular point deleted off.*

(2) Conversely, let G be a small finite subgroup of $GL(2, \mathbf{C})$ which is not a cyclic group. Then $T := \mathbf{C}^2/G-$ {the unique singular point} is a Platonic $\mathbf{C}^*$-fiber space with respect to the $\mathbf{C}^*$-action induced by the $\mathbf{C}^*$-action on $\mathbf{C}^2$ via the center of $GL(2, \mathbf{C})$. Moreover, the quotient morphism $q : \mathbf{C}^2 \to \mathbf{C}^2/G$ induces an étale finite Galois covering $\mathbf{C}^2 - (0) \to T$ with group G.

Remark that if $f : T \to C$ is a Platonic $\mathbf{C}^*$-fiber space and T is witten as $\mathbf{C}^2/G-\{\text{the unique singular point}\}$, we can construct a resolution $\tau : A \to \mathbf{C}^2/G$ (not necesarrily minimal), a completion $A \hookrightarrow V$ and a morphism $\pi : V \to C$ such that $\pi : V \to C$ is a normal completion of $f : T \to C$ and the exceptional curves of the resolution $\tau : A \to \mathbf{C}^2/G$ corresponds to one of the boundary components, say R_0.

For the structure of a Platonic $\mathbf{C}^*$-fiber space we have also the following:

Lemma 6 [**12**]. *Let G be a small finite subgroup of $GL(2, \mathbf{C})$, let $X := \mathbf{C}^2/G$ be a singular normal affine surface with G acting on $\mathbf{C}^2$ in a natural fashion via $GL(2, \mathbf{C})$, let P be the unique singular point of X which is the image of the origin of $\mathbf{C}^2$, and let $\theta : \mathbf{C}^2 \to X$ be the finite quotient morphism.*

Then the following three conditions are equivalent:

(1) *$X = \mathbf{C}^2/G$ is factorial;*
(2) *G is a binary icosahedral group in $SL(2, \mathbf{C})$;*
(3) *X is isomorphic to a hypersurface $x^2 + y^3 + z^5 = 0$ in $\mathbf{C}^3$.*

A platonic $\mathbf{C}^*$-fiber space is important in the theory of almost minimal models of noncomplete algebraic surfaces with logarithmic Kodaira dimension $-\infty$ and with non-connected boundaries at infinity. We recall the following result from [**10**].

Lemma 7 [**10**]. *Let X be a non-complete nonsingular algebraic surface defined over $\mathbf{C}$ with logarithmic Kodaira dimension $-\infty$. Suppose that X is not affine-ruled. Suppose, moreover, that there exists an open immersion of X into a nonsingular projective surface V such that*

(1) *$V - X$ is a reduced effective divisor with simple normal crossings;*
(2) *If we write $V - X = \bigcup_{i=1}^r C_i$ with irreducible components C_i, the intersection matrix $((C_i, C_j))_{1 \le i,j \le r}$ is not negative definite.*

Then there exists a Zariski open set U of X and a proper birational morphism $\phi : U \to T'$ onto a nonsingular algebraic surface T' defined over $\mathbf{C}$ such that

(a) *Either $U = X$ or $X - U$ has pure dimension one;*
(b) *T' is an open set of a Platonic $\mathbf{C}^*$-fiber space T with $\dim(T - T') \le 0$.*

We need one more technical lemma for the proof of Theorem 9.

Lemma 8 [**12**]. *Let $f : \mathbf{C}^n \to X$ be a dominant morphism to an irreducible affine surface defined over $\mathbf{C}$. Let $P_i (1 \le i \le r)$ be closed points of X such that either $f^{-1}(P_i) = \emptyset$ or $\dim f^{-1}(P_i) \le n - 2$ for every i. Then* *there exists a linear plane* L *of* $\mathbf{C}^n$ *such that* $f|_L\colon L \to X$ *is a dominant morphism and* $\dim f^{-1}(P_i) \cap L \leq 0$ *for every* i.

Now we will prove Theorem 9.

PROOF OF THEOREM 9: Let L be a linear plane of $\mathbf{C}^n$ such that $f|_L\colon \mathbf{C}^2 \to X$ is a dominant morphism and that either $f^{-1}(P) = \emptyset$ or $\dim f^{-1}(P) \cap L \leq 0$ for every singular point P of X. Such a linear plane L exists by virtue of Lemma 7.

Let $X' := X - \mathrm{Sing}X$ and $S' := L - f_L^{-1}(\mathrm{Sing}X)$. Since $f_L^{-1}(\mathrm{Sing}X)$ is a finite set of points (possibly empty), we know that $\kappa(S') = -\infty$. Therefore we have $\kappa(X') = -\infty$ by Lemma 2. On the other hand, we know that $\Gamma(X, \mathcal{O}_X)$ is factorial and $\Gamma(X, \mathcal{O}_X)^* = \mathbf{C}^*$. If X is nonsingular, then X is isomorphic to $\mathbf{C}^2$ by virtue of Theorems 1 and 2. Suppose that X is singular. If X' is affine ruled, X is isomorphic to $\mathbf{C}^2$ by Theorem 1. Hence X' is not affine ruled.

Then embed X into a normal projective surface W as an open subset in such a way that W is nonsingular along $W - X$ and $W - X$ is a reduced effective divisor with simple normal crossings. Let $\tilde{W} \to W$ be a desingularization of $\mathrm{Sing}(X)$ such that, with X' viewed as a Zariski open set of $\tilde{W}$, the complement $\tilde{W} - X'$ forms a reduced effective divisor with simple normal crossings. Since X is affine, the intersection matrix of $\tilde{W} - X'$ is not negative definite.

By Lemma 7, we know that X' contains an open subset U which satisfies the conditions of Lemma 7. But since X' is an open subset of an affine variety X and therefore X' does not contain complete curves, the proper birational morphism ϕ must be an isomorphism. Moreover from the proof of Lemma 7, we see that $T = T'$. Thus U itself is a Platonic $\mathbf{C}^*$-fiber space and $X' - U$ has pure dimension one. We claim that in fact $X' = U$. Assume the contrary. Then, since $\Gamma(X, \mathcal{O}_X)$ is factorial, there exists an element b of $\Gamma(X', \mathcal{O}_{X'})$ such that $X' - U$ is the support of the zero locus of b, i.e., $\mathrm{Supp}(b)_{0,X'} = \mathrm{Supp}(X' - U)$. Hence b is invertible on U and $b \notin \mathbf{C}^*$. On the other hand, a Platonic $\mathbf{C}^*$-fiber space U can be written as $\mathbf{C}^2/G-$ {the unique singular point}. Therefore b can be extended to an invertible function $\tilde{b}$ on $\mathbf{C}^2/G$ which induces an invertible function on $\mathbf{C}^2$. Therefore $b \in \mathbf{C}^*$, which is a contradiction.

Since X is normal and $\dim(X - X') \leq 0$, we have

$$\Gamma(X, \mathcal{O}_X) = \Gamma(X', \mathcal{O}_{X'}) = \Gamma(U, \mathcal{O}_U) = \Gamma(\mathbf{C}^2/G, \mathcal{O}_{\mathbf{C}^2/G}).$$

By the hypothesis that $\Gamma(X, \mathcal{O}_X)$ is factorial, we conclude that $X = \mathbf{C}^2/G$ is isomorphic to the hypersurface $x^2 + y^3 + z^5 = 0$ in $\mathbf{C}^3$ by Lemma 6.

Now we prove Theorem 8.

PROOF OF THEOREM 8: (1) Let $X := \mathrm{Spec}A = \mathrm{Spec}\mathbf{C}[X_1, \cdots, X_n]^G$ and let $f : \mathbf{C}^n \to X$ be a quotient morphism. Then X is a normal affine surface defined over $\mathbf{C}$. First we show that A is factorial. Since $A^* = (\mathbf{C}[X_1, \cdots, X_n])^* = \mathbf{C}^*$, it is enough to prove that every irreducible element of A is also irreducible as an element of $\mathbf{C}[X_1, \cdots, X_n]$. Let $a \in A$ and assume that a is written as a product of a constant c and irreducible elements $z_1, \cdots, z_r$ of $\mathbf{C}[X_1, \cdots, X_n]$ as follows:

$$a = cz_1^{m_1} \cdots z_r^{m_r}$$

where $m_1, \cdots, m_r \in \mathbf{N}$.

Since a group G acts on R, we have

$$a = {}^g a = c({}^g z_1)^{m_1} \cdots ({}^g z_r)^{m_r},$$

for $g \in G$. From the uniqueness of the representation of a as a product of irreducible elements of R, we have

$${}^g z_i = \chi_i(g) \cdot z_{\alpha_g(i)}$$

for some map $\chi_i(g) : G \to \mathbf{C}^*$ and $\alpha_g : \{1, \cdots, r\} \to \{1, \cdots, r\}$. Then α_g induces a representation of G on the permutation group of r letters. But since G is a connected algebraic group, this representation must be trivial. Therefore we have

$${}^g z_i = \chi_i(g) \cdot z_i$$

and $\chi : G \to \mathbf{C}^*$ is a multiplicative character. By the assumption on G, χ_i is trivial and z_i is invariant under the action of G. Thus $z_i \in A$ and A is factorial.

(2) Next we prove that for any point P of X either $f^{-1}(P) = \emptyset$ or $\dim f^{-1}(P) \leq n - 2$. Assume that there exists an irreducible component Z of dimension $n-1$ of $f^{-1}(P)$ for some point $P \in X$. Then since R is factorial, there exists an element t of R whose zero locus is Z. For an element g of G, we have $f({}^g Z) = P = f(Z)$ and ${}^g Z \subset f^{-1}(P)$. Since a number of irreducible components of $f^{-1}(P)$ of dimension $n-1$ is finite, we have ${}^g Z = Z$ for $\forall g \in G$ by the same reason as in (1).

Therefore we have

$${}^g t = \xi(g) \cdot t, \ \xi(g) \in \mathbf{C}^*.$$

Since $\xi : G \to \mathbf{C}^*$ is a multiplicative character, $\xi(g) = 1$ for all $g \in G$ by the assumption. Thus we have ${}^g t = t$ and $t \in A$. This is a contradiction.

(3) Now apply Theorem 9 to the present morphism $f : \mathbf{C}^n \to X$. We have $X \simeq \mathbf{C}^2$ or X is isomorphic to the hypersurface $x^2 + y^3 + z^5 = 0$ in $\mathbf{C}^3$. Assume that $X \not\simeq \mathbf{C}^2$. Let P be the unique singular point of X and let $\pi : Z \to X - \{P\}$ be the topological universal covering. Then the quotient morphism $f : \mathbf{C}^n \to X$ induces:

$$f' : \mathbf{C}^n - f^{-1}(P) \longrightarrow Z \longrightarrow X - \{P\},$$

where f' is the restriction of f onto $\mathbf{C}^n - f^{-1}(P)$. Hence general fibers of f' are reducible. However, by the construction of an affine quotient variety X, general fibers of f must be irreducible and reduced. This is a contradiction.

Finally to prove Theorem 4, we need one more lemma from [**13**].

Lemma 4.5 [**13**]. *Let $f : X \to Y$ be a dominant morphism from an affine variety $X = \mathrm{Spec}(A)$ of dimension $n+1$ to a smooth affine variety $Y = \mathrm{Spec}(B)$ of dimension n. Assume that*

(1) *general fibers of f are isomorphic to $\mathbf{A}^1_{\mathbf{C}}$,*
(2) *both A and B are factorial,*
(3) $A^* = \mathbf{C}^*$,
(4) *f is geometrically irreducible in codimension 1 over Y.*

Then there exists a closed subset Z of Y such that

(a) *if $Z \neq \emptyset$, every irreducible component of Z has codmension 2 in Y,*
(b) *$U := Y - Z$ coincides with $f(X)$, and $f : X \to U$ is an $\mathbf{A}^1_{\mathbf{C}}$-bundle.*

Now we prove Theorem 4.

Proof of Theorem 4: Let A be an affine domain over $\mathbf{C}$ of dimension three and let $X = \mathrm{Spec}(A)$ be as in the statement of Theorem 4. Since X is affine 1-ruled, G_a acts on A by Lemma 1. Put $S = A^{G_a}, Y = \mathrm{Spec}(S)$ and let $\pi : X \to Y$ be a quotient morphism. Then Y is isomorphic to X/G_a. We define an action of G_a^{n+1} on $X \times \mathbf{A}^n$ as follows:

$$(t_0, t_1, \cdots, t_n)(x, (s_1, \cdots, s_n)) = (t_0 \cdot x, t_1 + s_1, \cdots, t_n + s_n)$$

for $x \in X, t_0, \cdots, t_n, s_1, \cdots, s_n \in \mathbf{C}$. The quotient of $X \times \mathbf{A}^n$ by this action is Y. On the other hand, since $X \times \mathbf{A}^n \simeq \mathbf{A}^{n+3}$, we have $X \times \mathbf{A}^n / G_a^{n+1} \simeq \mathbf{A}^2$ by Theorem 8. Therefore $Y \simeq \mathbf{A}^2$.

Clealy a general fiber of π is isomorphic to $\mathbf{A}^1_{\mathbf{C}}$.We define U as in Lemma 9. Then we have

$$1 = e(X) = e(U) = 1 - \#(Y - U).$$

This means $U = Y \simeq \mathbf{A}^2$ and we conclude that $X \simeq \mathbf{A}^3$.

REMARK: We note that three conditions B-1), B-2) and C-2) are not sufficient to characterize the affine three space $\mathbf{A}^3$. Let S be a homology plane, that is, let S be a nonsingular affine surface whose homology group $H_i(S, \mathbf{Z})$ is trivial for $i > 0$ and $S \neq \mathbf{A}^2$. See [**3**], [**14**] and [**16**] for an example of homology plane. Put $X = S \times \mathbf{A}^1_{\mathbf{C}}$. Then it is easy to see that X satisfies conditions B-1), B-2) and C-2). See [**2**]. But X is not isomorphic to $\mathbf{A}^3$ by Cancellation Theorem for $\mathbf{A}^2$.

References

[**1**] T.Fujita, *On Zariski problem*, Proc. Japan Acad., Ser. A **55** (1979), 106-110.

[**2**] T. Fujita, *On the topology of non-complete algebraic surfaces*, J. Fac. Sci. Univ. Tokyo, Sect 1A **29** (1982), 503-566.

[**3**] R.V.Gurjar and M.Miyanishi, *Affine surfaces with $\bar{\kappa} \leq 1$*, Algebaic geometry and commutative algebra in honor of Masayoshi Nagata (1987), 99-124.

[**4**] R.V.Gurjar and A.R. Shastri, *The fundamental group at infinity of an affine surface*, Comm. Math. Helv. **59** (1984), 459-484.

[**5**] S.Iitaka, *On logarithmic Kodaira dimension of algebraic varieties*, Complex analysis and algebraic geometry, A collection of papers dedicated to K.Kodaira (1977), 175-189.

[**6**] M.Miyanishi, *An algebraic characterization of the affine plane*, J. Math. Kyoto Univ. **15** (1975), 169-184.

[**7**] M.Miyanishi and T.Sugie, *Affine surfaces containing cylinderlike open set*, J. Math. Kyoto Univ. **20** (1980), 11-42.

[**8**] M.Miyanishi, *Regular subrings of a polynomial ring*, Osaka J. Math. **17** (1980), 329-338.

[**9**] M.Miyanishi, *Non-complete algebraic surfaces*, Lecture Notes in Math., 857, Springer, Berin- Heidelberg-NewYork, (1981).

[**10**] M.Miyanishi and S. Tsunoda, *Non-complete algebraic surfaces with logarithmic Kodaira dimension $-\infty$ and with non-connected boundaries at infinity,*, Japan J. Math. **10** (1984), 195-242.

[**11**] M.Miyanishi and S.Tsunoda, *Logarithmic del Pezzo surfaces of rank one with noncontractible boundaries*, Japan J. Math. **10** (1984), 271-319.

[**12**] M.Miyanishi, *Normal affine subalgebras of a polynomial ring*, Algebraic and Topological Theories-to the memory of Dr. Takehiko Miyata, Kinokuniya, Tokyo (1985), 37-51.

[**13**] M.Miyanishi, *Algebraic characterization of the affine 3-space*, Proc. Algebraic Geometry Seminar(Singapore), (1988), 53-67.

[**14**] M.Miyanishi and T.Sugie, *Examples of homology planes of general type*, preprint.

[**15**] D.Mumford, *Enriques' classification of surfaces in char. p, 1*, Global analysis. (1969).

[**16**] C.P. Ramanujam, *A topological characterization of the affine plane as an algebraic variety*, Ann. of Math. **94** (1971), 69-88.

Toru Sugie
Department of Mathematics
Kyoto University
Kyoto
JAPAN

CLASSIFICATION OF THREE-DIMENSIONAL HOMOGENEOUS COMPLEX MANIFOLDS

JÖRG WINKELMANN

Introduction

A complex manifold X is called homogeneous if there exists a connected complex or real Lie group G acting transitively on X as a group of biholomorphic transformations. The goal is a general classification of homogeneous complex manifolds. Since the class of homogeneous complex manifolds is much too big for any serious attempt of complete classification, it is necessary to impose further conditions. For example É. Cartan classified in **[Ca]** symmetric homogeneous domains in $\mathbb{C}^n$. Here we will require that X is of small dimension. For $dim_{\mathbb{C}}(X) = 1$ the classification follows from the uniformization Theorem. In 1962 J. Tits classified the compact homogeneous complex manifolds in dimension two and three **[Ti1]**. In 1979 J. Snow classified all homogeneous manifolds $X = G/H$ with $dim_{\mathbb{C}}(X) \leq 3$, G being a solvable complex Lie group and H discrete **[SJ1]**. The classification of all complex-homogeneous (i.e. G is a complex Lie group) two-dimensional manifolds was completed in 1981 by A. Huckleberry and E. Livorni **[HL]**. Next, in 1984 K. Oeljeklaus and W. Richthofer classified all those homogeneous two-dimensional complex manifolds $X = G/H$ where G is only a real Lie group **[OR]**. The classification of three-dimensional complex-homogeneous manifolds was completed in 1985 **[W1]**. Finally in 1987 the general classification of the three-dimensional homogeneous complex manifolds was given by our Dissertation **[W2]**. The purpose of this note is to describe these manifolds and briefly outline the methods involved in the classification.

One should note that our classification does not cover all complex manifolds on which $Aut_{\mathcal{O}}(X)$ acts transitively. For example on $X = \mathbb{C}^2 \setminus \{(0,0),(0,1)\}$ no real or complex Lie group acts transitively by biholomorphic transformations, although the whole group of automorphisms

$Aut_{\mathcal{O}}(X)$ *does* act transitively. (The first example of this kind is due to Kaup [**K**]).

The complete List

The following list covers all homogeneous complex manifolds $X = G/H$ with $dim_{\mathbb{C}}(X) \leq 3$:

We distinguish the cases G solvable, G mixed and G semisimple. Here G is mixed means that G has a Levi-Malcev decomposition $G = S \ltimes R$ with $dim_{\mathbb{R}}(R) > 0$ and $dim_{\mathbb{R}}(S) > 0$, i.e. G is neither semisimple nor solvable.

G COMPLEX SOLVABLE

(1) Quotients G/Γ of solvable complex Lie groups G with $dim_{\mathbb{C}}(G) \leq 3$ by discrete subgroups.

This class contains in particular $\mathbb{C}^n$, $\mathbb{C}^*$ *and Tori. These manifolds have been studied in detail in* [**SJ1**]. *She gives a fine classification of the discrete subgroups of these solvable Lie groups.*

G COMPLEX SEMISIMPLE

(2) Quotients $SL_2(\mathbb{C})/\Gamma$ with Γ being a discrete subgroup of $SL_2(\mathbb{C})$.

This is a very large class. For example let M *be an arbitrary Riemann surface. Then there is a holomorphic action of* $\pi_1(M)$ *on the universal covering* $\tilde{M}$ *of* M. *Since* $\tilde{M} \simeq \mathbb{P}_1$, $\mathbb{C}$, *or* Δ_1, *the universal covering* $\tilde{M}$ *is equivariantly embeddable in* $\mathbb{P}_1$. *Thus for any Riemann surface the fundamental group* $\pi_1(M)$ *can be embedded in* $SL_2(\mathbb{C})/\mathbb{Z}_2 \simeq Aut_{\mathcal{O}}(\mathbb{P}_1)$ *as a discrete subgroup.*

(3) The following homogeneous-rational manifolds:
 a) $\mathbb{P}_n$ for $n \leq 3$,
 b) the projective quadric $\mathbf{Q}_3$ and
 c) the flag manifold $F_{1,2}(3)$ of full flags in $\mathbb{C}^3$.

(4) The affine quadric Q_2 and $\mathbb{P}_2 \setminus \mathbf{Q}_1$.

Both are quotients of $SL_2(\mathbb{C})$ *by reductive subgroups and* $\mathbb{P}_2 \setminus \mathbf{Q}_1 \simeq Q_2/\mathbb{Z}_2$. *Furthermore* Q_2 *is biholomorphic to* $\{(z,w) \in \mathbb{P}_1 \times \mathbb{P}_1 \mid z \neq w\}$ *and may be realized as affine bundle over* $\mathbb{P}_1$. *In contrast* $\mathbb{P}_2 \setminus \mathbf{Q}_1$ *has no equivariant fibration at all.*

(5) All $\mathbb{C}^*$– and Torus-principal bundles over homogeneous rational manifolds.

This class contains in particular $\mathbb{C}^2 \setminus \{(0,0)\}$, $\mathbb{C}^3 \setminus \{(0,0,0)\}$, *homogeneous Hopf surfaces and* $\mathbb{P}_3 \setminus (L_1 \dot{\cup} L_2)$ *where* L_1 *and* L_2 *are two disjoint complex lines in* $\mathbb{P}_3$.

G COMPLEX MIXED

(6) The non-trivial $\mathbb{C}^*$– and torus-principal bundles over $\mathbb{P}_2 \setminus \mathbf{Q}_1$.

The non-trivial $\mathbb{C}^*$*– and torus-principal bundles over* Q_2 *are also homogeneous manifolds, but are already contained in the class* $SL_2(\mathbb{C})/\Gamma$.

(7) Every line bundle over a homogeneous-rational manifold which is generated by a positive divisor.

This class contains in particular $\mathbb{P}_n \setminus \{x_0\}$.

(8) Holomorphic vector bundles of rank two over $\mathbb{P}_1$ which are direct sums of line bundles generated by positive divisors.

Any vector bundle of rank two over $\mathbb{P}_1$ *is a direct sum of line bundles (see* [**GrR**, p.237]*), but of course not necessarily generated by positive divisors.*

The total space of the vector bundle $E \simeq H^1 \oplus H^1$ *is* $\mathbb{P}_3 \setminus L$, *where* L *denotes a complex line in* $\mathbb{P}_3$. *Furthermore* E *may be realized as a* $\mathbb{C}$*-principal bundle over* H^2. *Here* H^2 *denotes the* 2nd *power of the hyperplane bundle over* $\mathbb{P}_1$.

(9) Quotients of $\mathbb{P}_3 \setminus L$ realized as principal bundle over H^2 by discrete subgroups of the structure group.

It is easy to list all these quotient, since it suffices to determine the discrete subgroups of the one-dimensional structure group.

(10) Every line bundle over Q_2 and the unique non-trivial line bundle over $\mathbb{P}_2 \setminus \mathbf{Q}_1$.

(11) Quotients of $\mathbb{C} \times Q_2$ by discrete subgroups of $\mathbb{Z}_2 \ltimes (\mathbb{C}, +)$ with the $\mathbb{Z}_2 \ltimes \mathbb{C}$–action on $\mathbb{C} \times Q_2$ given by

$$([z],[w],y) \mapsto ([z],[w],y+x)$$

for $(e,x) \in \mathbb{Z}_2 \ltimes \mathbb{C}$ and

$$([z],[w],y) \mapsto ([w],[z],-y)$$

for $(\phi, 0)$, where ϕ denotes the non-trivial element of $\mathbb{Z}_2$. (Here $([z],[w]) \in \mathbb{P}_1 \times \mathbb{P}_1 \setminus \Delta \simeq Q_2$).

(12) Quotients of $\mathbb{C} \times (\mathbb{C}^2 \setminus \{(0,0)\})$ by discrete subgroups of $\mathbb{C}^* \ltimes \mathbb{C}$ acting by

$$(\lambda, z) : (x, v) \mapsto (\lambda^k x + z, \lambda v)$$

for $k \in \mathbb{Z}$.

(13) Certain $\mathbb{C}^2$–bundles over $\mathbb{P}_1$ which are given by the following transition functions

$$w_1 = -\left(\frac{z_0}{z_1}\right)^n w_0$$
$$v_1 = \left(\frac{z_0}{z_1}\right)^{np+n-2} v_0 - \left(\frac{z_0}{z_1}\right)^{np+n-1} w_0^{p+1}$$

for $p \geq 1$, $n \geq 1$.

Here v_i and w_i denote fibre coordinates over $U_i = \{[z_0 : z_1] \mid z_i \neq 0\}$.

These bundles arise as quotients of $SL_2(\mathbb{C}) \ltimes N$ by a three-codimensional subgroup where N is a complex nilpotent Lie group with $dim_{\mathbb{C}}(N/N') = n+1$ and $N^{(p)} \neq \{e\} = N^{(p+1)}$. The commutator N' is abelian and induces a fibration which realizes these manifolds as affine bundles over H^n where H^n denotes the n–th power of the hyperplane bundle over $\mathbb{P}_1$. These affine bundles have no holomorphic section and the manifolds have only constant holomorphic functions.

For $n = 1$ and $p = 1$ the group N is the three-dimensional complex Heisenberg group, i.e.

$$N = \left\{ \begin{pmatrix} 1 & x & y \\ & 1 & z \\ & & 1 \end{pmatrix} \middle| \; x, y, z \in \mathbb{C} \right\},$$

and the affine bundle over $H^n = H^1$ is actually a principal bundle. Moreover for $n = p = 1$ the manifold which arises is biholomorphic to $\mathbf{Q}_3 \setminus L$, where $\mathbf{Q}_3$ denotes the projective quadric and L an arbitrary complex line in $\mathbf{Q}_3$.

(14) Quotients E/Γ where E is the $\mathbb{C}$–principal bundle over H^1 which is contained in the above class for $n = p = 1$ and Γ is a discrete subgroup of the structure group $(\mathbb{C}, +)$ acting from the right on E.

(15) Simply-connected $\mathbb{C}^*$–principal bundles over H^1 which are given as a quotient $(SL_2(\mathbb{C}) \ltimes N)/H$ where N is the three-dimensional complex Heisenberg group, the representation of $SL_2(\mathbb{C})$ in $Aut(N/N')$ is irreducible and

$$H = \left\{ \left(\begin{pmatrix} e^z & w \\ & e^{-z} \end{pmatrix}; \begin{pmatrix} 1 & x & \alpha z \\ & 1 & \\ & & 1 \end{pmatrix} \right) \middle| \, x, z, w \in \mathbb{C} \right\}$$

for $\alpha \in \mathbb{C}^*$.

(16) Quotients of the above principal bundles by discrete subgroups of the principal structure group acting from the right.

G REAL SOLVABLE

(17) An irreducible bounded homogeneous domain, i.e. a ball

$$\mathbb{B}_n = \{(z_1, \dots, z_n) \in \mathbb{C}^n \mid \sum_{i=1}^{n} |z_i|^2 < 1\}$$

$$\simeq \{(z_1, \dots, z_n) \in \mathbb{C}^n \mid \sum_{i=2}^{n} |z_i|^2 < Re(z_1)\}$$

and

$$\Omega = \{(x, w, z) \in \mathbb{C}^3 \mid Im\, x > 0 \quad \text{and} \quad 4\, Im\, x\, Im\, z > (Im\, w)^2\}.$$

For $dim_{\mathbb{C}}(X) \leq 3$ every bounded homogeneous domain is also a hermitian symmetric space. In the notation of **[Hel]**, *$\mathbb{B}_n$ is a hermitian domain of type $A\,III(p = 1, q = n)$ and Ω is of type $BD\,I(p = 3, q = 2) = C\,I(n = 2)$.*

(18) A complement to a bounded domain in its equivariant embedding in $\mathbb{C}^n$, i.e.

$$X \simeq \{(z_1, \dots, z_n) \in \mathbb{C}^n \mid \sum_{i=2}^{n} |z_i|^2 > Re(z_1)\}$$

or

$$X \simeq \{(x, w, z) \in \mathbb{C}^3 \mid Im\, x > 0 \quad \text{and} \quad 4\, Im\, x\, Im\, z < (Im\, w)^2\}.$$

(19) $\mathbb{C}^2\backslash\mathbb{R}^2$ or a covering of this manifold.

The manifold $\mathbb{C}^2\backslash\mathbb{R}^2$ *is not simply-connected,* $\mathbb{C}^2\backslash\mathbb{R}^2 \simeq S^1 \times \mathbb{R}^3$. *The universal covering* $\widetilde{\mathbb{C}^2\backslash\mathbb{R}^2}$ *which is diffeomorphic to* $\mathbb{R}^4$ *has some interesting complex-analytic properties. In particular* $\widetilde{\mathbb{C}^2\backslash\mathbb{R}^2}$ *is hypersurface-separable (i.e. for all* $x, y \in \widetilde{\mathbb{C}^2\backslash\mathbb{R}^2}$ *there exists a hyperplane* $H \subset \widetilde{\mathbb{C}^2\backslash\mathbb{R}^2}$ *such that* $x \in H \not\ni y$*) but it is not meromorphically separable. Actually any meromorphic function on* $\widetilde{\mathbb{C}^2\backslash\mathbb{R}^2}$ *is* $\pi_1(\mathbb{C}^2\backslash\mathbb{R}^2)$*-invariant. Hence two points in the same fiber over* $\mathbb{C}^2\backslash\mathbb{R}^2$ *can not be separated.*

(20) A quotient of $\mathbb{C} \times \widetilde{\mathbb{C}^2\backslash\mathbb{R}^2}$ by a discrete subgroup of $\mathbb{C} \times \mathbb{Z}$ acting naturally ($\mathbb{C}$ on $\mathbb{C}$ by translations and $\mathbb{Z} \simeq \pi_1(\mathbb{C}^2\backslash\mathbb{R}^2)$ on $\widetilde{\mathbb{C}^2\backslash\mathbb{R}^2}$ as a group of covering transformations).

(21) The following domains in $\mathbb{C}^3$

$$\begin{aligned}
\Omega_0 &= \{(x, w, z) \mid Im\, x > 0 \quad \text{and} \quad f_1(x, w, z) > 0\} \\
\Omega_1 &= \{(x, w, z) \mid f_1(x, w, z) > 0\} \\
\Omega_2 &= \{(x, w, z) \mid f_2(x, w, z) > 0\} \\
\Omega_3 &= \{(x, w, z) \mid f_2(x, w, z) < 0\}
\end{aligned}$$

with $f_1 = Im\, z - Re\, w\, Im\, x$ and $f_2 = Im\, z - Re\, w\, Im\, x + (Re\, x)^4$.

The manifold Ω_0 *is particular interesting for its Kobayashi-reduction.*

The Kobayashi-reduction identifies two points in a manifold if their Kobayashi-pseudometric is zero (for details about the Kobayashi-pseudometric see [**Ko1, Ko3**]*). Now the Kobayashi-reduction of* Ω_0 *is a fibration*

$$\pi : G/H \xrightarrow{\Delta_1 \times \mathbb{C}} G/I \simeq \Delta_1$$

compatible with the complex structure. In particular the fiber has a non-trivial Kobayashi-pseudometric. Nevertheless if one takes any open subset U *of* G/I *then the Kobayashi-pseudometric of* $\pi^{-1}(U)$ *degenerates along the fibers.*

One can define a "complex-line-reduction" for Ω_0 *which identifies two points* $x, y \in \Omega_0$ *if and only if there is a finite chain of holomorphic maps* $\phi_1, \ldots \phi_n : \mathbb{C} \to \Omega_0$ *with* $\phi_0(0) = x$, $\phi_i(1) = \phi_{i+1}(0)$ *and* $\phi_n(1) = y$. *Then* $\Omega_0 \to \Omega_0/\sim$ *is a* G*-equivariant real analytic fiber bundle and all the fibers are closed complex-analytic subsets of* Ω_0 *but there is no compatible complex structure on* $\Omega_0/\sim$.

That Ω_2 and Ω_3 are not biholomorphic is proved in the following way: Assume to the contrary that $\phi : \Omega_2 \to \Omega_3$ is a biholomorphic map. Obviously ϕ is extendable to the envelopes of holomorphy i.e. to the whole $\mathbb{C}^3$. Then $-f_2 \circ \phi$ and f_2 must define the same boundary. Hence $-f_2 \circ \phi = \lambda f_2$ for some positive real-analytic function λ. One obtains a contradiction by writing down this equation in coordinates and comparing the coefficients of the power series up to degree 4.

The automorphism groups of all these homogeneous open domains in $\mathbb{C}^3$ are finite-dimensional Lie groups (see [**Tan**]*).*

For Ω_0 and Ω_1 there exist real solvable Lie groups acting freely and transitively. In contrast to this, any Lie group G which acts transitively on Ω_2 or Ω_3 is at least real seven-dimensional.

G REAL NON-SOLVABLE

(22) The manifold $\{(x,w,z) \in \mathbb{C}^3 \mid Im\, z - Re\, w\, Im\, x > 0,\ (x,w) \notin \mathbb{C}^2 \backslash \mathbb{R}^2\}$ and coverings of this manifold.

This is homogeneous under a non-trivial semidirect product of $GL_2(\mathbb{R})$ and the real three-dimensional Heisenberg group.

(23) $\mathbb{C}^3 \setminus \mathbb{R}^3$

This manifold is homogeneous under an $SL_3(\mathbb{R}) \ltimes (\mathbb{R}^3,+)$-action. It is diffeomorphic to $S^2 \times \mathbb{R}^4$, hence simply-connected.

(24) $\{(x,w,z) \in \mathbb{C}^3 \mid 4\, Im\, x\, Im\, z - (Im\, w)^2 < 0\}$ and coverings of this manifold.

This manifold is diffeomorphic to $S^1 \times \mathbb{R}^5$. It is one of three open orbits of a $GL_2(\mathbb{R}) \ltimes (\mathbb{R}^3,+)$-action on $\mathbb{C}^3$ (The other two orbits are bounded domains).

(25) The line bundles over $\mathbb{P}_2(\mathbb{C}) \setminus \mathbb{P}_2(\mathbb{R})$, $\mathbb{P}_2 \setminus \overline{\mathbb{B}_2}$ and $\mathbb{P}_1 \times \mathbb{P}_1 \setminus \mathbb{P}_1^R$ which are generated by the restriction of a positive divisor on $\mathbb{P}_2$ resp. $\mathbb{P}_1 \times \mathbb{P}_1$. Here $\mathbb{P}_1^R$ means $\{[x],[w] \in \mathbb{P}_1 \times \mathbb{P}_1 \mid [\bar{x}] = [w]\}$.

These manifolds are homogeneous under an $S \ltimes (\Gamma(X,L),+)$-action where S is a real semisimple group acting transitively on the base X and $\Gamma(X,L)$ is the finite-dimensional additive group of sections in the line bundle.

(26) Those $\mathbb{C}^*$– and torus-principal bundles over $\mathbb{P}_2(\mathbb{C}) \setminus \mathbb{P}_2(\mathbb{R})$ and $\mathbb{P}_2 \setminus \overline{\mathbb{B}_2}$ which are extendable to principal bundles over $\mathbb{P}_2$ resp. $\mathbb{P}_1 \times \mathbb{P}_1$.

The group $S \times P$ acts transitively on these manifolds, where S is a real semisimple group acting transitively on the base and P the principal structure group acting from the right.

(27) Simply-connected principal bundles over $\mathbb{P}_1 \times \mathbb{P}_1 \setminus \mathbb{P}_1^R$ for any $[\lambda : \mu] \in \mathbb{P}_1(\mathbb{C})$ arising as quotients $X_{[\lambda:\mu]} = G/(G \cap \hat{H}_{[\lambda:\mu]})^0$ with

$$G = \{(z, A, B) \in \mathbb{C} \times SL_2(\mathbb{C}) \times SL_2(\mathbb{C}) \mid A = \bar{B}\}$$

and

$$\hat{H}_{[\lambda:\mu]} = \{(\lambda x + \mu y, \begin{pmatrix} e^x & \\ w & e^{-x} \end{pmatrix}, \begin{pmatrix} e^y & z \\ & e^{-y} \end{pmatrix} \mid x, y, z, w \in \mathbb{C}\}$$

Here the complex structure on G/H comes from the quotient vector space $\hat{\mathfrak{s}}/\hat{\mathfrak{h}}_{[\lambda:\mu]}$ to which G/H is locally isomorphic.

(28) Quotients of the above listed simply-connected principal bundles by discrete subgroups of the principal structure group acting from the right.

Let $X = G/H = G/(G \cap \hat{H}_{[\lambda:\mu]})^0$ denote the above listed simply-connected principal bundles. Let $H_1 := G \cap \hat{H}_{[\lambda:\mu]}$. Denote the G-anti-canonical fibration (see [**HO**]*) of X by $G/H \to G/I$.*

The subgroup

$$S = SL_2(\mathbb{C})^R := \{(A, \bar{A}) \in SL_2(\mathbb{C}) \times SL_2(\mathbb{C}) \mid A \in SL_2(\mathbb{C})\}$$

of G acts transitively on G/H iff $|\lambda| = |\mu|$. Hence for $|\lambda| = |\mu|$ the manifold $X_{[\lambda:\mu]}$ is biholomorphic to the Lie group $SL_2(\mathbb{C})$ equipped with a left-invariant complex structure other than the usual one.

The group $\hat{H}_{[\lambda:\mu]}$ is a closed Lie subgroup of

$$\hat{G} = \mathbb{C} \times SL_2(\mathbb{C}) \times SL_2(\mathbb{C})$$

iff

$$[\lambda : \mu] \in (\mathbb{P}_1(\mathbb{C}) \setminus \mathbb{P}_1(\mathbb{R})) \cup \mathbb{P}_1(\mathbb{Q}).$$

If $\hat{H}_{[\lambda:\mu]}$ is closed, then $G/H_1 \to G/I$ is the restriction of a principal bundle

$$\hat{G}/\hat{H} \to \hat{G}/\hat{I} \simeq \mathbb{P}_1 \times \mathbb{P}_1$$

with fiber

$$\hat{I}/\hat{H} \simeq \mathbb{C}/ < \lambda, \mu >_{\mathbb{Z}} .$$

In particular $\hat{I}/\hat{H} \simeq \mathbb{C}^*$ *if* $[\lambda:\mu] \in \mathbb{P}_1(\mathbb{Q})$.

$$H_1/H = \pi_0(G/H_1) \simeq \begin{cases} \mathbb{Z} & \text{if } [\lambda:\mu] \notin \mathbb{P}_1(\mathbb{Q}) \text{ or } \lambda = -\mu \\ \mathbb{Z}_{|r+s|} & \text{if } [\lambda:\mu] = [r:s] \text{ with } r,s \in \mathbb{Z}, \\ & \gcd(r,s) = 1 \text{ and } r \neq -s \ . \end{cases}$$

Therefore G/H_1 *is simply-connected only if* $[\lambda:\mu] = [r:1-r]$ *with* $r \in \mathbb{Z}$.
The G*–anticanonical fibration* $G/H \to G/I$ *has the fiber*

$$I/H \simeq \begin{cases} \mathbb{C}^* & \text{if } \lambda \neq -\mu \\ \mathbb{C} & \text{if } \lambda = -\mu \end{cases}$$

Open orbits of real forms in homogeneous-rational manifolds

(29) $\mathbb{P}_n(\mathbb{C}) \setminus \mathbb{P}_n(\mathbb{R})$ with $n = 2$ or $n = 3$.

These manifolds are homogeneous under an $SL_{n+1}(\mathbb{R})$*–action. They are strictly pseudoconcave.*

(30) $\Omega^+_{n,m} = \{[z] \in \mathbb{P}_{n+m-1}(\mathbb{C}) \mid \|z\|^2_{n,m} > 0\}$ for $1 \leq n,m$ and $n+m \leq 4$.

Here

$$\|z\|^2_{n,m} = \sum_{i=0}^{n-1} |z_i|^2 - \sum_{i=n}^{n+m-1} |z_i|^2$$

for $z = (z_0, \ldots, z_{n+m-1})$. *Note that* $\mathbb{B}_m \simeq \Omega^+_{1,m}$ *is a bounded domain and that* $\mathbb{P}_3 \setminus \overline{\mathbb{B}_n} \simeq \Omega^+_{n,1}$ *is a strictly pseudoconcave manifold. The unit disk* $\mathbb{B}_1 \simeq \Omega^+_{1,1}$ *is biholomorphic to both of the connected components of* $\mathbb{P}_1(\mathbb{C}) \setminus \mathbb{P}_1(\mathbb{R})$. $\Omega^+_{1,3}$ *and* $\Omega^+_{3,1}$ *are homogeneous under an* $SU_{1,3}$*–action,* $\Omega^+_{2,2}$ *is homogeneous under* $SU_{2,2}$, $Sp_{1,1}$ *and* $Sp_2(\mathbb{R})$.

(31) $F_R = \{(z,w) \in \mathbb{P}_2 \times \mathbb{P}_2 \mid z^t w = 0, z, w \notin \mathbb{P}_2(\mathbb{R})\}$.

This manifold is homogeneous under $SL_3(\mathbb{R})$ *acting by*

$$(A) : (z,w) \mapsto (A^t z, A^{-1} w).$$

(32) $F^+_{+,+} = \{(z,w) \in \mathbb{P}_2 \times \mathbb{P}_2 \mid z^t w = 0, \|z\|^2_{2,1} > 0, \|w\|^2_{2,1} < 0\}$ and
$F^+_{+,-} = \{(z,w) \in \mathbb{P}_2 \times \mathbb{P}_2 \mid z^t w = 0, \|z\|^2_{2,1} > 0, \|w\|^2_{2,1} > 0\}$.

The group $SU_{2,1}$ *acts transitively on these two manifolds by*

$$A : (z,w) \mapsto (A^t z, A^{-1} w).$$

(33) $Q^+_{n,5-n} = \mathbb{Q}_3 \cap \Omega^+_{n,5-n} = \{[z] \in \mathbb{P}_4 \mid z^t z = 0, \|z\|^2_{n,5-n} > 0\}$ for $2 \leq n \leq 4$.

$SO(n,m)$ *acts transitively.* $Q^+_{2,3}$ *is a bounded domain and* $Q^+_{4,1}$ *is strictly pseudoconcave.*

(34) $\mathbb{P}_1 \times \mathbb{P}_1 \setminus \mathbb{P}_1^{\mathbb{R}} = \{(z,w) \in \mathbb{P}_1 \times \mathbb{P}_1 \mid \bar{z} \neq w\}$.

This is a strictly pseudoconcave manifold homogeneous under an action of $SL_2(\mathbb{C})$ as a real Lie group. This action is given by $A : (z,w) \mapsto (Az, \bar{A}w)$.

DIRECT PRODUCTS

(35) Direct products of the above listed manifolds.

Further results

BOUNDED DOMAINS

Our classification implicitly contains a new proof of the fact that every three-dimensional bounded homogeneous domain is a symmetric space. This fact was first stated in 1935 by È. Cartan in [**Ca**]. The proof was omitted, because he considered its length to be out of proportion to the interest of the result. Cartan also conjectured in this article that in general every bounded homogeneous domain is a symmetric space. But in 1959 Pyatetski-Shapiro detected a four-dimensional counterexample to this general conjecture (see [**PS1**]).

Furthermore we obtained the following result, which is true in arbitrary dimensions:

Proposition. *Let $G/H \to G/I$ be a fibration of homogeneous complex manifolds with both fiber and base biholomorphic to a bounded homogeneous domain. Assume furthermore that the fiber I/H is a symmetric space. Then the total space G/H is also a bounded domain.*

The methods of the classification

We classify the three-dimensional connected homogeneous complex manifolds in the following way.

We use both the classification of homogeneous surfaces ([**HL, OR, Hu**]) and the classification of three-dimensional complex-homogeneous manifolds in [**W**].

The classification of the three-dimensional complex-homogeneous manifolds, i.e. the manifolds for which there exists a complex Lie group acting holomorphically and transitively is based on a variety of methods. A key point in all these classifications is the Levi-Malcev decomposition. If G is any simply-connected real or complex Lie group then G is a semidirect product $G = S \ltimes R$ of the radical R and a maximal connected semisimple

Lie subgroup S. Hence there are three possibilities for a homogeneous complex manifold X: Either there exist a solvable Lie group acting transitively or there is a semisimple Lie group which acts transitively or every Lie group G acting transitively on X has a Levi-Malcev decomposition $G = S \ltimes R$ with both $dim_{\mathbb{R}}(S) > 0$ and $dim_{\mathbb{R}}(R) > 0$.

THE CASE G COMPLEX AND SOLVABLE

The classification of the three-dimensional complex-homogeneous manifolds starts with the case that a solvable complex Lie group acts transitively on X. In this case we proved in **[W]** the following

Proposition. *Let G be a solvable, connected complex Lie group which acts transitively on a complex manifold X with $dim_{\mathbb{C}}(X) \leq 3$. Then there exists another solvable connected complex Lie group G^* acting transitively on X with $dim_{\mathbb{C}}(X) = dim_{\mathbb{C}}(G^*)$.*

This reduces this case to the situation where $X = G/H$, G is a three-dimensional solvable complex Lie group and H a discrete subgroup, a situation studied in detail in **[SJ1]** by J. Snow.

To prove this proposition we begin by considering the Tits-fibration. For a complex-homogeneous manifold G/H this is a holomorphic fiber bundle $G/H \to G/N_G(H^0)$ with parallelizable fiber $N_G(H^0)/H$ and a base $G/N_G(H^0)$ which is equivariantly embeddable in some $\mathbb{P}_N$ (see **[HO]**). Here $N_G(H^0)$ denotes the normalizer of the connected component H^0 of H in G. Actually $\pi(x) = \pi(y)$ for any $x, y \in X$ is equivalent to x, y having the same isotropy algebra. This is in fact the reason why the base is equivariantly embeddable in a projective space, because one can think of the Tits-fibration as a map from X into the Grassmannian manifold of $dim_{\mathbb{C}}(H)$–dimensional vector subspaces of the Lie algebra $\mathbf{g}$ given by $x \mapsto \mathbf{h}_x$. In the solvable case the fiber is homogeneous under an action of the solvable Lie group N/H^0 (where N is an abbreviation for $N_G(H^0)$). and the base G/N is biholomorphic to $\mathbb{C}^n \times (\mathbb{C}^*)^k$ **[SD]**, i.e. as a manifold, the base is biholomorphic to an abelian complex Lie group A. The main idea now is to construct a new group G^* from the Lie groups A and N/H^0. We may assume that G is simply-connected. Hence G/H^0 and G/N^0 are the universal coverings of G/H and G/N, and $G/N \simeq \mathbb{C}^n \times (\mathbb{C}^*)^k$ implies that $N/N^0 \simeq \mathbb{Z}^k$. It follows that N/H^0 equals $(\ldots(N^0/H^0 \ltimes \mathbb{Z})\ldots \ltimes \mathbb{Z})$. From explicit knowledge of all one- and two-dimensional simply-connected solvable Lie groups (there are only three) we can deduce that all these semidirect products are extendable in such a way that N/H^0 is embedded in some

$$G^* := (\ldots(N^0/H^0 \ltimes \mathbb{C})\ldots \ltimes \mathbb{C})$$

with $G^*/(N^0/H^0) \simeq \tilde{A} \simeq (\mathbb{C}^{k+n}, +)$. Now G^* contains H/H^0 as a discrete subgroup and is as a manifold biholomorphic to a product of N^0/H^0 and G/N^0. Therefore any section in the trivial holomorphic principal bundle

$$\mathbb{C}^3 \simeq G/H^0 \rightarrow G/N^0 \simeq \mathbb{C}^{k+n}$$

induces a biholomorphic map from G^* to G/H^0 which is N^0/H^0–equivariant. It remains to show that this map can be chosen such that it is H/H^0–equivariant. This problem is equivalent to solving a certain functional equation which in turn boils down to a $\bar{\partial}$–problem on the Stein manifold $G/N^0 \simeq \mathbb{C}^{k+n}$. Thus the proof is completed.

THE CASE G COMPLEX AND NON-SOLVABLE

The classification for the case that G is semisimple is not too difficult, because there exists a complete classification of all semisimple Lie groups. Moreover there are classifications of maximal Lie subgroups of semisimple Lie groups (see [**Dy1,Dy2**]), from which it follows that most semisimple Lie groups can't act non-trivially on a three-dimensional manifold. In fact, if S is a simply-connected semisimple complex Lie group acting transitively on a three-dimensional manifold then $S \simeq Sp_2(\mathbb{C})$, $S \simeq SL_n(\mathbb{C})$ with $n \leq 4$, $S \simeq SL_2(\mathbb{C}) \times SL_k(\mathbb{C})$ with $k \leq 3$ or $S \simeq SL_2(\mathbb{C}) \times SL_2(\mathbb{C}) \times SL_2(\mathbb{C})$.

The case that G is neither solvable nor semisimple is more interesting. If $dim_{\mathbb{C}}(S) > 3$ it is easy to deduce that X is a line bundle over $\mathbb{P}_2$ or $\mathbb{P}_1 \times \mathbb{P}_1$. Thus the remaining case is $G \simeq SL_2(\mathbb{C}) \ltimes R$. Key points in the classification of this case are the discussion of fibrations induced by the radical R and its commutator R' and in particular the use of representation theory to understand the representation of $S = SL_2(\mathbb{C})$ in $Aut(R)$. This finally leads to a number of homogeneous $\mathbb{C}^2$–bundles over $\mathbb{P}_1$ and some other homogeneous manifolds.

G REAL

Let us now consider the case where G is only a real Lie group. This weakened assumption causes some new difficulties. For example the structure of fibrations $G/H \rightarrow G/I$ is much more complicated. In the complex-homogeneous setting these fibrations are always holomorphic fiber bundles. But as we will see below, if G is only a real Lie group a fiber bundle $G/H \rightarrow G/I$ with compatible G-left-invariant complex structures on G/H and G/I need not be locally holomorphically trivial. Furthermore, if G/H is a complex manifold and $G/H \rightarrow G/K$ is a real fibration, where all the fibres are closed complex analytic subsets of G/H, it is still possible that there exists no compatible G-left-invariant complex structure on G/K.

HOLOMORPHIC FIBRATIONS

Even if G is only a real Lie group there are some fibrations which yield holomorphic fiber bundles. The most important such fibration is the anticanonical fibration. This is a generalization of the Tits-fibration $G/H \to G/N_G(H^0)$. The problem in our situation is that the normalizer $N_G(H^0)$ of H^0 is not related to the complex structure on X. Therefore $G/N_G(H^0)$ has in general no compatible complex structure and may even be real odd-dimensional.

Hence we define the anticanonical fibration in general as follows. Note that there is a canonical Lie algebra homomorphism ϕ of $\mathbf{g}$ into the complex Lie algebra $\Gamma_{\mathcal{O}}(X, T_{\mathcal{O}}X)$ of global holomorphic vector fields on X. Let $\mathbf{g}^C$ denote the complexification of the real Lie algebra $\phi(\mathbf{g})$ in $\Gamma_{\mathcal{O}}(X, T_{\mathcal{O}}X)$. Define the "complex isotropy algebra" $\mathbf{h}_x^C$ at a point $x \in X$ by $\mathbf{h}_x^C = \{\mathbf{X} \in \mathbf{g}^C \mid \mathbf{X}|_x = 0\}$. Then the anticanonical fibration $X \to X/\sim$ defined by

$$x \sim y \iff \mathbf{h}_x^C = \mathbf{h}_y^C$$

maps X into a complex Grassmannian manifold. Defined in this way the anticanonical fibration is furthermore always a holomorphic fibre bundle with a complex-homogeneous, parallelizable fiber.

Somewhat generalizing this anticanonical fibration we will prove the following result

Proposition. *Let G/H be a homogeneous complex manifold of arbitrary dimension. Then there exists a fibration $G/H \to G/J$ with the following properties:*

(i) The map $G/H \to G/J$ is a holomorphic fibre bundle.

(ii) The base G/J is equivariantly embeddable in some $\mathbb{P}_N$. Furthermore the anticanonical fibration of G/J is injective.

(iii) The representation of the Lie algebra $\mathbf{g}$ in $\Gamma(G/J, T_{\mathcal{O}}(G/J))$ is totally real.

Now this fibration has the convenient property that it is a holomorphic fibre bundle, i.e. it has a local holomorphic trivialization, but nevertheless if the base G/J has bad complex-analytic properties then it might be very difficult to classify such holomorphic fibre bundles. E.g. $H^1(X, \mathcal{O})$, the moduli space of $\mathbb{C}$–principal bundles over X is infinite-dimensional and rather complicated for some homogeneous surfaces like $\mathbb{C}^2 \setminus \mathbb{R}^2$. Therefore we try to extend the fiber bundle $\pi : G/H \to G/J$ to a J/H–fiber bundle over $\hat{G}/\hat{J}$, where $\hat{G}/\hat{J}$ is the orbit in $\mathbb{P}_N$ of the complexification $\hat{G}$ of G in $Aut_{\mathcal{O}}(\mathbb{P}_N) \simeq PSL_{N+1}(\mathbb{C})$. This is desirable, because $\hat{G}/\hat{J}$ has in general nice complex-analytic properties, e.g. if G is solvable then $\hat{G}/\hat{J} \simeq \mathbb{C}^k \times (\mathbb{C}^*)^n$ and hence $H^1(\hat{G}/\hat{J}, \mathcal{O}) = 0$.

It turns out that this may be achieved in most cases in the following way: Let $\mathbf{g}^C$ denote the complexification of the Lie algebra $\mathbf{g}$ in $\Gamma_{\mathcal{O}}(X, T_{\mathcal{O}}X)$ and $\mathbf{h}^C$ the "complex isotropy" defined as above. Define G^C as the simply-connected complex Lie group corresponding to $\mathbf{g}^C$ and H^C as the connected Lie subgroup of G^C corresponding to $\mathbf{h}^C$. If H^C is closed in G^C then the manifold $\hat{X} = G^C/H^C$ is a principal bundle over $\hat{G}/\hat{J}$ which is almost an extension of the bundle $G/H \to G/J$.

If $G \cap H^C = H$ then we obtain the following commutative diagram

$$\begin{array}{ccc} G/H & \hookrightarrow & G^C/H^C \\ \downarrow & & \downarrow \\ G/J & \hookrightarrow & \hat{G}/\hat{J} \hookrightarrow \mathbb{P}_N. \end{array}$$

(In general instead of $G \cap H^C = H$ we obtain only $(G \cap H^C)^0 = H^0$, but this causes only minor technical difficulties.)

The only case in which H^C may be not closed is the case $dim_{\mathbb{C}}(\hat{S}) \geq 6$, where $\hat{S}$ denotes the maximal semisimple subgroup of $\hat{G}$. But in this special case the assumption $dim_{\mathbb{C}}(G/H) \leq 3$ is strong enough to determine the structure of $\hat{G}$ and G/H in detail.

With these methods we prove the classification for the case that the "maximal holomorphic fibration" $G/H \to G/J$ has positive-dimensional fibers.

From then on we can assume that the fibration $G/H \to G/J$ has discrete fibers. This implies in particular that the anticanonical fibration of G/H has discrete fibers and that the representation of the Lie algebra $\mathbf{g}$ in $\Gamma_{\mathcal{O}}(X, T_{\mathcal{O}}X)$ is totally real.

Moreover we can use the embedding $G/J \hookrightarrow \hat{G}/\hat{J} \hookrightarrow \mathbb{P}_N$ to obtain new fibrations which are compatible with the complex structure on G/J. Let $\hat{I}$ denote any closed complex Lie subgroup of $\hat{G}$ containing $\hat{J}$. Then we have the following diagram of fibrations

$$\begin{array}{ccccc} G/H & & & & \\ \downarrow & & & & \\ G/J & \hookrightarrow & \hat{G}/\hat{J} & \hookrightarrow & \mathbb{P}_N \\ \downarrow \pi & & \downarrow \hat{\pi} & & \\ G/I = G/(G \cap \hat{I}) & \hookrightarrow & \hat{G}/\hat{I} & & \end{array}$$

Caveat: The bundle $\hat{\pi} : \hat{G}/\hat{J} \to \hat{G}/\hat{I}$ is always a holomorphic fiber bundle, but only in the case $I/J = \hat{I}/\hat{J}$ does this imply that $\pi : G/I \to G/J$ is also a holomorphic fiber bundle.

THE CASE G REAL SOLVABLE

In the solvable situation our next steps consist of efforts to ensure that the $dim_{\mathbb{R}}(G)$ is not too big if we assume that there is no group of smaller dimension acting transitively on G/H. Ideally we would like to have $dim_{\mathbb{R}}(G) = dim_{\mathbb{R}}(G/H)$. The following is the main tool for this purpose.

Proposition. *Let G be a real Lie group (not necessarily solvable) acting holomorphically on a complex manifold G/H and $G/H \to G/I$ be a fibration compatible with the complex structure. Assume that no complex Lie group acts non-trivially on I/H. Assume that the G–action on G/H is almost effective and denote by L the ineffectivity of G on G/I. Then L acts almost effectively on any fiber of $G/H \to G/I$.*

In other words if a vector field $\mathbf{X} \in \mathbf{g}$ vanishes on the base and on one fibre, it must vanish on all fibres. This proposition is a very useful tool not only in the solvable situation. In the solvable case it is the main step in the proof that, with only two exceptions, we may assume $dim_{\mathbb{R}}(G) = dim_{\mathbb{R}}(G/H) = 6$. One can not apply it when the fiber I/H equals $\mathbb{C} \times H^+$. But in this case it is possible to prove that either $dim_{\mathbb{R}}(G) = dim_{\mathbb{R}}(G/H)$ or G/H is trivial or a $\mathbb{C} \times H^+$–bundle over a homogeneous Riemann surface with some special properties. In the first case, i.e. $dim_{\mathbb{R}}(G) = dim_{\mathbb{R}}(G/H)$, we continue with a discussion of the nilradical N of G. We note that

$$dim_{\mathbb{R}}(N) \geq \frac{1}{2} dim_{\mathbb{R}}(G)$$

holds for any solvable Lie group G. Hence $3 \leq dim_{\mathbb{R}}(N) \leq 6$ in our situation. The case $dim_{\mathbb{R}}(N) = 6$, i.e. $N = G$, can be omitted, because we prove that under this assumption $X = G/H$ is a complex-homogeneous manifold:

Proposition. *Let G/H be any homogeneous complex manifold, where G is a real nilpotent Lie group. Then there exists a transitive action of G^C on G/H, where G^C is the abstract complexification of G.*

For $3 \leq dim_{\mathbb{R}}(N) \leq 5$ it is not too difficult to determine all possible nilpotent Lie groups N. We always assume that G/H is "non-trivial", i.e. neither complex-homogeneous nor a direct product. From these assumptions we deduce that if N is abelian then $dim_{\mathbb{R}}(N) = 3$. Furthermore it follows that if $dim_{\mathbb{R}}(N) = 5$ then $dim_{\mathbb{R}}(Z) = 1$, where Z denotes the center of N. Finally if $dim_{\mathbb{R}}(N) = \frac{1}{2} dim_{\mathbb{R}}(G)$ then N is necessarily abelian. This leaves only six possibilities for the nilradical N.

Now it remains to determine the Lie group structure of G and the leftinvariant complex structure on G. Observe that in general a solvable Lie group doesn't carry any left-invariant complex structure at all. For example consider the group $G = (\mathbb{R}, +) \ltimes_\rho (\mathbb{R}^3, +)$ defined by the group homomorphism

$$\rho(t) = \begin{pmatrix} e^{\alpha t} & & \\ & e^{\beta t} & \\ & & e^{\gamma t} \end{pmatrix}$$

On this group there exists a left-invariant complex structure if and only if at least two of the values α, β, γ coincide.

Furthermore, even if there exists a left-invariant structure, it needn't be an interesting one. For example, since our goal is to classify homogeneous complex manifold we are not interested in making a long list of real solvable Lie groups with left-invariant structures such that G as a manifold is biholomorphic to $\mathbb{C}^3$.

Hence it wouldn't be very efficient to start classifying all real six-dimensional solvable Lie groups and then classify the left-invariant structures on them. Instead, our strategy is the following: For every possible nilradical N we try to determine the group structure and the left-invariant complex structure of G simultaneously under permanent exploitation of our assumptions that G as a manifold is neither a direct product nor homogeneous under a complex Lie group.

We start by choosing a real vector space base for the Lie algebra $\mathbf{g}$ as canonically as possible. If possible, we choose an element $\mathbf{Z}$ in a real one-dimensional ideal of $\mathbf{g}$ (Actually this is always possible in our situation). From the assumption that G is not biholomorphic to a direct product we deduce that $\mathbf{Z}$ does not commute with $J\mathbf{Z}$. Hence wlog $[J\mathbf{Z}, \mathbf{Z}] = \mathbf{Z}$ and in particular $J\mathbf{Z} \notin \mathbf{n}$. Next we define

$$\mathbf{c} := \{\mathbf{X} \in \mathbf{g} \mid [\mathbf{X}, \mathbf{Z}] = 0\}.$$

Then

$$\mathbf{g} = < J\mathbf{Z}, \mathbf{Z} >_{\mathbb{R}} \oplus (\mathbf{c} \cap J\mathbf{c}).$$

Now we discuss the $ad(J\mathbf{Z})$–action on $J\mathbf{c} \cap \mathbf{c}$. From the integrability condition it follows that

$$[J\mathbf{Z}, J\mathbf{C}] = J[J\mathbf{Z}, \mathbf{C}]$$

for all $\mathbf{C} \in J\mathbf{c} \cap \mathbf{c}$. Hence $J\mathbf{c} \cap \mathbf{c}$ is $ad(J\mathbf{Z})$–stable.

In generalizing the eigenspace decomposition for not necessarily diagonalizable endomorphisms we define weight spaces V_λ. Let V be a real vector space, ϕ an endomorphism, and

$$V_\lambda^C := \{x \in V^C = V \otimes_{\mathbb{R}} \mathbb{C} \mid \exists N : (\phi - \lambda id)^N(x) = 0\}.$$

Then

$$V = \bigoplus_{Im\,\lambda \geq 0} (V \cap (V_\lambda^C + V_{\bar{\lambda}}^C)).$$

Thus we obtain a complete decomposition of V. Now assume that there is a Lie algebra structure on V and that ϕ is a derivation. Then

$$[V_\lambda^C, V_\mu^C] \subset V_{\lambda+\mu}.$$

We use this to analyse the $ad(J\mathbf{Z})$–action on $J\mathbf{c} \cap \mathbf{c}$.

These methods finally yield a complete description of the Lie algebra $\mathbf{g}$ and its J–structure. This determines the complex structure on G. To realize this homogeneous manifold in $\mathbb{C}^3$, we then give a concrete realization of this Lie algebra as an algebra of globally integrable holomorphic vector fields on $\mathbb{C}^3$. This yields a G–action on $\mathbb{C}^3$. A given point in $\mathbb{C}^3$ lies in an open orbit of G iff the $\mathbf{g}$–vector fields span the whole tangent space at this point. Thus it is possible to determine the open orbits by explicit calculations with determinants.

G REAL SOLVABLE, $dim_{\mathbb{R}}(G) > 6$

In the solvable situation we still have to discuss the subcase that no real six-dimensional Lie group acts transitively. This is possible only under special circumstances. In particular this implies that there is a fibration

$$G/H \xrightarrow{H^+ \times \mathbb{C}} G/J.$$

A key point in the further analysis is the group $A := (N \cap J)^0$. We prove that this group A is actually an abelian normal Lie subgroup of G. Furthermore $\mathbf{g} = <\mathbf{A}_1, \mathbf{A}_2, \mathbf{A}_3>_{\mathbb{R}} \oplus \mathbf{a}$ for appropriately chosen $\mathbf{A}_i \in \mathbf{g}$. Now the $ad(\mathbf{A}_i)$–actions must stabilize the totally real subspace $\mathbf{a}$ in $\hat{\mathbf{a}} = \mathbf{a} + i\mathbf{a}$. From this we finally deduce that $dim_{\mathbb{R}}(A) \leq 4$. Since $dim_{\mathbb{R}}(G) > 6$ this implies $dim_{\mathbb{R}}(A) = 4$. Using explicit calculations with vector fields on $\mathbb{C}^3$ we then classify the possible manifolds, which are only the following two.

$$\Omega^+ = \{(x, w, z) \mid Im\,z - Re\,w\,Im\,x + (Re\,x)^4 > 0\}$$
$$\Omega^- = \{(x, w, z) \mid Im\,z - Re\,w\,Im\,x + (Re\,x)^4 < 0\}$$

G MIXED

Next let us discuss the case where G is neither solvable nor semisimple, i.e. $G = S \ltimes R$ with $dim_{\mathbb{R}}(R) > 0$, $dim_{\mathbb{R}}(S) > 0$. The nilpotent normal Lie subgroup $A := (G' \cap R)^0$ plays a key role in this case. Since we now may assume that the above described "maximal holomorphic" fibration is injective, the center of G is discrete. This implies that R is not central in G, hence $dim_{\mathbb{R}}(A) > 0$. Furthermore $G/H \hookrightarrow \hat{G}/\hat{H} \hookrightarrow \mathbb{P}_N$.

Observe that $\hat{A}'$ can't act transitively on positive-dimensional $\hat{A}$-orbits (This is in fact true for any nilpotent group $\hat{A}$). From this we can deduce that either there exist fibrations

$$\begin{array}{ccc} G/H & \hookrightarrow & \hat{G}/\hat{H} \\ \downarrow & & \downarrow \\ G/I & \hookrightarrow & \hat{G}/\hat{I} \end{array}$$

or the group $\hat{A}$ is abelian and acts transitively on $\hat{G}/\hat{H}$. In the latter case we obtain $\hat{G} = \hat{H} \ltimes \hat{A}$ with $\hat{A} = (\mathbb{C}^3, +)$. Consequently $\hat{G} \subset GL_3(\mathbb{C}) \ltimes (\mathbb{C}^3, +)$ and $G \subset GL_3(\mathbb{R}) \ltimes (\mathbb{R}^3, +)$. We then continue with the discussion of the maximal semisimple Lie subgroup S in G and its representation in $GL_3(\mathbb{R})$.

On the other hand, if $\hat{A}$ is not abelian or not acting transitively then $\hat{I} := \hat{A}'\hat{H}$ (resp. $\hat{I} := \hat{A}\hat{H}$) yields proper fibrations. Here we again use the above mentioned result on the ineffectivity L of the G–action on G/I. In particular we exploit the fact that L acts almost effectively on I/H if $I/H \simeq \mathbb{C}^2 \backslash \mathbb{R}^2$ or $I/H \simeq H^+$. This helps in controlling $dim_{\mathbb{R}}(G)$.

Observe that there aren't many homogeneous manifolds of dimension one or two on which a semisimple group can act. This is in particular true, if there is simultaneously a solvable group acting on this manifold. The most important such surface is $\mathbb{C}^2 \backslash \mathbb{R}^2$ with either $SL_2(\mathbb{R}) \ltimes \mathbb{R}^2$ or $GL_2(\mathbb{R}) \ltimes \mathbb{R}^2$ acting.

The maximal semisimple group S of G is a real form of $\hat{S}$, since we may here assume that G is totally real. Furthermore it must stabilizes a totally real subspace $\mathbf{r}$. This allows further conclusions, because in general if $\hat{S}$ is a complex semisimple group, $\rho : \hat{S} \to GL_{\mathbb{C}}(V^C)$ a representation, and S a real form $\hat{S}$, there doesn't necessarily exist any S–stable totally real subspace at all. In particular we prove that if $\hat{S} \simeq SL_2(\mathbb{C})$, $S \simeq SU_2$ and ρ irreducible, then a totally real S–stable subspace can exist only if V^C is odd-dimensional. Together with the fact that $dim_{\mathbb{R}}(N/N')$ is even-dimensional for any Heisenberg group N, this is a key argument in the classification in this case.

G REAL SEMISIMPLE

The last case is the case where G is semisimple and the "maximal holomorphic" fibration almost injective. Now G is totally real and hence $dim_{\mathbb{C}}(\hat{G}) \geq 6$. Therefore $\hat{G}/\hat{H}$ is a three-dimensional homogeneous manifold with a transitive action of a complex semisimple Lie group, which is at least six-dimensional. Furthermore the anticanonical fibration is injective. This implies that $\hat{G}/\hat{H}$ is one of the following three manifolds: $\mathbb{P}_3$, the projective quadric Q_3 or the flag manifold $F_{1,2}(3)$. In particular $\hat{G}/\hat{H}$ is homogeneous-rational and we are in a situation studied in general by Wolf (see [**Wo**]). Explicit calculations then yield the classification.

References

[**Bo**] Borel, A.: Linear algebraic groups. New York. Benjamin (1969)

[**Ca**] Cartan, E.: Sur les domaines bornés homogènes de l'espace de n variables complexes. *Abh. Math. Sem. Hamburg* **11/12**, 116-162 (1935)

[**Dy1**] Dynkin, E.B.: Semisimple subalgebras of semisimple Lie algebras. *Mat. Sbornik N. S.* **30** (72), 349-462, Moskva (1952)

[**Dy2**] Dynkin, E.B.: Maximal subgroups of classical groups. *Trudy Moskov. Mat. Obšč.1*, **39-66**, Moskva (1952)

[**Gi**] Gilligan, B.: On bounded holomorphic reductions of homogeneous spaces. *CR Math. Rep. Acad. Sci. Canada*, **6** no.4, 175-178 (1984)

[**Go**] Goto, M.: Faithful representations of Lie groups I. *Math. Japonicae* **1**, 107-119 (1948).

[**Gr**] Grauert, H.: Analytische Faserungen über holomorph–vollständigen Räumen. *Math. Ann.* **135**, 263-273 (1958)

[**He**] Helgason, S.: Differential Geometry and Symmetric Spaces. New York: Academic Press (1962)

[**Hu**] Huckleberry, A.T.: The classification of homogeneous surfaces. *Expo. Math.* **4** (1986), 189-334

[**HL**] Huckleberry, A. T., Livorni, E. L.: A classification of homogeneous surfaces. *Can. J. Math.* **33**, 1097-1110 (1981)

[**HO1**] Huckleberry, A. T.; Oeljeklaus, E.: Classification Theorems for almost homogeneous spaces. Pub. Inst. E. Cartan **9** (1984)

[**HO2**] Huckleberry, A. T.; Oeljeklaus, E.: Homogeneous spaces from a complex analytic viewpoint. Manifolds and Lie groups. (Papers in honor of Y. Matsushima), Progress in Math., Birkhäuser, Boston (1981)

[**HS**] Huckleberry, A. T.; Snow, D.: A classification of strictly pseudoconcave homogeneous manifolds. *Ann. Scuola. Norm. Sup. Pisa, Série IV,Vol.* **VIII**, 231-255 (1981)

[**K**] Kaup, W.: Reelle Transformationsgruppen und invariante Metriken auf komplexen Räumen. *Inventiones Math.* **3**, 43-70 (1967).

[Kr] Kraft, H.: Geometrische Methoden in der Invariantentheorie. Aspekte der Mathematik, Band D1. Vieweg, Braunschweig 1984

[Ko1] Kobayashi, S.: Hyperbolic manifolds and holomorphic mappings. Dekker Inc., New York 1970

[Ko2] Kobayashi, S.: Intrinsic distances, measures and geometric function theory. *Bull. AMS* **82** no.3, 357-416 (1976)

[Na] Nakajima, K.: Hyperbolic homogeneous manifolds and Siegel domains. *J. Math. Univ. Kyoto* **25**, 269 - 291 (1985)

[OR] Oeljeklaus, K.; Richthofer, W.: Homogeneous complex surfaces. *Math. Ann.* **268**, 273-292 (1984)

[PS] Pyatetskii-Shapiro, I. I.: The Geometry and classification of bounded homogeneous domains. *Russ. Math. Surveys* **20**, p.1 (1965)

[PS1] Pyatetskii-Shapiro, I. I.: On a problem proposed by É. Cartan. *Dokl. Akad. Nauk SSSR* **124** (1959), 272-273

[SD] Snow, D.: Stein quotients of connected complex Lie groups. *Manuscripta Math.* **50**, 185–214 (1985)

[SJ1] Snow, J.: Complex solv-manifolds of dimension two and three. Thesis, Notre Dame Univ., Indiana (1979)

[SJ2] Snow, J.: On the classification of solv-manifolds in dimension two and three. Revue de l'Institut E. Cartan **10**, Nancy (1986)

[Tan] Tanaka, N.: On the pseudoconformal geometry of hypersurfaces of the space of n complex variables. *J. Math. Soc. Japan* **14**, 397-429 (1962)

[Ti] Tits, J.: Espaces homogènes complexes compacts. *Comm. Math. Helv.* **37**, 111-120 (1962)

[W1] Winkelmann, J.: Klassifikation dreidimensionaler komplex-homogener Mannigfaltigkeiten. Diplomarbeit. Ruhr-Universität Bochum (1985)

[W2] Winkelmann, J.: The classification of three-dimensional homogeneous complex manifolds. Dissertation. Ruhr-Universität Bochum (1987)

[Wo] Wolf, J. A.: The action of a real semisimple group on a complex flag manifold. I: orbit structures and holomorphic arc components. *Bull. A.M.S.* **75**, 1121-1237 (1969)

Jörg Winkelmann
Mathematisches Institut NA 4/75
Ruhr-Universität Bochum
Universitätsstraße 150
D-4630 Bochum 1
WEST-GERMANY

Progress in Mathematics

Edited by:

J. Oesterlé
Departement des Mathematiques
Université de Paris VI
4, Place Jussieu
75230 Paris Cedex 05
France

A. Weinstein
Department of Mathematics
University of California
Berkeley, CA 94720
U.S.A.

Progress in Mathematics is a series of books intended for professional mathematicians and scientists, encompassing all areas of pure mathematics. This distinguished series, which began in 1979, includes authored monographs and edited collections of papers on important research developments as well as expositions of particular subject areas.

All books in the series are "camera-ready", that is they are photographically reproduced and printed directly from a final-edited manuscript that has been prepared by the author. Manuscripts should be no less than 100 and preferably no more than 500 pages.

Proposals should be sent directly to the editors or to: Birkhäuser Boston, 675 Massachusetts Avenue, Suite 601, Cambridge, MA 02139, U.S.A.

1 GROSS. Quadratic Forms in Infinite-Dimensional Vector Spaces
2 PHAM. Singularités des Systèmes Différentiels de Gauss-Manin
3 OKONEK/SCHNEIDER/SPINDLER. Vector Bundles on Complex Projective Spaces
4 AUPETIT. Complex Approximation. Proceedings, Quebec, Canada, July 3–8, 1978
5 HELGASON. The Radon Transform
6 LION/VERGNE. The Weil Representation. Maslov Index and Theta Series
7 HIRSCHOWITZ. Vector Bundles and Differential Equations Proceedings, Nice, France, June 12–17, 1979
8 GUCKENHEIMER/MOSER/NEWHOUSE. Dynamical Systems, C.I.M.E. Lectures, Bressanone, Italy, June 1978
9 SPRINGER. Linear Algebraic Groups
10 KATOK. Ergodic Theory and Dynamical Systems I
11 BALSLEV. 18th Scandinavian Congress of Mathematicians, Aarhus, Denmark, 1980
12 BERTIN. Séminaire de Théorie des Nombres, Paris 1979–80
13 HELGASON. Topics in Harmonic Analysis on Homogeneous Spaces
14 HANO/MARIMOTO/MURAKAMI/OKAMOTO/OZEKI. Manifolds and Lie Groups: Papers in Honor of Yozo Matsushima
15 VOGAN. Representations of Real Reductive Lie Groups
16 GRIFFITHS/MORGAN. Rational Homotopy Theory and Differential Forms
17 VOVSI. Triangular Products of Group Representations and Their Applications
18 FRESNEL/VAN DER PUT. Géometrie Analytique Rigide et Applications
19 ODA. Periods of Hilbert Modular Surfaces
20 STEVENS. Arithmetic on Modular Curves
21 KATOK. Ergodic Theory and Dynamical Systems II
22 BERTIN. Séminaire de Théorie des Nombres. Paris 1980–81
23 WEIL. Adeles and Algebraic Groups

24 Le Barz/Hervier. Enumerative Geometry and Classical Algebraic Geometry
25 Griffiths. Exterior Differential Systems and the Calculus of Variations
26 Koblitz. Number Theory Related to Fermat's Last Theorem
27 Brockett/Millman/Sussman. Differential Geometric Control Theory
28 Mumford. Tata Lectures on Theta I
29 Friedman/Morrison. Birational Geometry of Degenerations
30 Yano/Kon. CR Submanifolds of Kaehlerian and Sasakian Manifolds
31 Bertrand/Waldschmidt. Approximations Diophantiennes et Nombres Transcendants
32 Books/Gray/Reinhart. Differential Geometry
33 Zuily. Uniqueness and Non-Uniqueness in the Cauchy Problem
34 Kashiwara. Systems of Microdifferential Equations
35 Artin/Tate. Arithmetic and Geometry: Papers Dedicated to I.R. Shafarevich on the Occasion of His Sixtieth Birthday, Vol. 1
36 Artin/Tate. Arithmetic and Geometry: Papers Dedicated to I.R. Shafarevich on the Occasion of His Sixtieth Birthday. Vol. II
37 de Monvel. Mathématique et Physique
38 Bertin. Séminaire de Théorie des Nombres, Paris 1981–82
39 Ueno. Classification of Algebraic and Analytic Manifolds
40 Trombi. Representation Theory of Reductive Groups
41 Stanely. Combinatories and Commutative Algebra
42 Jouanolou. Théorèmes de Bertini et Applications
43 Mumford. Tata Lectures on Theta II
44 Kac. Infinite Dimensional Lie Algebras
45 Bismut. Large Deviations and the Malliavin Calculus
46 Satake/Morita. Automorphic Forms of Several Variables Taniguchi Symposium, Katata, 1983
47 Tate. Les Conjectures de Stark sur les Fonctions L d'Artin en s = 0
48 Fröhlich. Classgroups and Hermitian Modules
49 Schlichtkrull. Hyperfunctions and Harmonic Analysis on Symmetric Spaces
50 Borel, et al. Intersection Cohomology
51 Bertin/Goldstein. Séminaire de Théoire des Nombres. Paris 1982–83
52 Gasqui/Goldschmidt. Déformations Infinitesimales des Structures Conformes Plates
53 Laurent. Théorie de la Deuxième Microlocalisation dans le Domaine Complèxe
54 Verdier/Le Potier. Module des Fibres Stables sur les Courbes Algébriques Notes de l'Ecole Normale Supérieure, Printemps, 1983
55 Eichler/Zagier. The Theory of Jacobi Forms
56 Shiffman/Sommese. Vanishing Theorems on Complex Manifolds
57 Riesel. Prime Numbers and Computer Methods for Factorization
58 Helffer/Nourrigat. Hypoellipticité Maximale pour des Opérateurs Polynomes de Champs de Vecteurs
59 Goldstein. Séminarie de Théorie des Nombres, Paris 1983–84
60 Procesi. Geometry Today: Giornate Di Geometria, Roma. 1984
61 Ballmann/Gromov/Schroeder. Manifolds of Nonpositive Curvature
62 Guillou/Marin. A la Recherche de la Topologie Perdue
63 Goldstein. Séminaire de Théorie des Nombres, Paris 1984–85
64 Myung. Malcev-Admissible Algebras
65 Grubb. Functional Calculus of Pseudo-Differential Boundary Problems

66 Cassou-Noguès/Taylor. Elliptic Functions and Rings and Integers

67 Howe. Discrete Groups in Geometry and Analysis: Papers in Honor of G.D., Mostow on His Sixtieth Birthday

68 Robert. Antour de L'Approximation Semi-Classique

69 Faraut/Harzallah. Deux Cours d'Analyse

70 Adolphson/Conrey/Ghosh/Yager. Number Theory and Diophantine. Problems: Proceedings of a Conference at Oklahoma State University

71 Goldstein. Séminaire de Théories des Nombres, Paris 1985–1986

72 Vaisman. Symplectic Geometry and Secondary Characteristics Classes

73 Molino. Reimannian Foliations

74 Henkin/Leiterer. Andreotti–Grauert Theory by Integral Formulas

75 Goldstein, Séminaire de Théories des Nombres, Paris 1986–87

76 Cossec/Dolgachev. Enriques Surfaces I

77 Reysaat. Quelques Aspects des Surfaces de Riemann

78 Borho/Brylinski/MacPherson. Nilpotent Orbits, Primitive Ideals, and Characteristic Classes

79 McKenzie/Valeriote. The Structure of Decidable Locally Finite Varieties

80 Kraft/Petrie/Schwarz. Topological Methods in Algebraic Transformation Groups